Computational Fluid Mechanics

Selected Papers

Computational Fluid Mechanics

Selected Papers

Alexandre Joel Chorin

Department of Mathematics
University of California
Berkeley, California

ACADEMIC PRESS, INC.

Harcourt Brace Jovanovich, Publishers

Boston San Diego New York
Berkeley London Sydney
Tokyo Toronto

ACADEMIC PRESS, INC.
1250 Sixth Avenue, San Diego, CA 92101

United Kingdom Edition published by
ACADEMIC PRESS INC. (LONDON) LTD.
24-28 Oval Road, London NW1 7DX

Library of Congress Cataloging-in-Publication Data

Chorin, Alexandre Joel.
Computational fluid mechanics : selected papers / Alexandre Joel Chorin.
p. cm.
Includes bibliographical references.
ISBN 0-12-174070-6 (alk. paper)
1. Fluid dynamics. 2. Numerical analysis. I. Title.
QA911.C445 1989
532'pr.05—dc20 89-17551
CIP

Printed in the United States of America
89 90 91 92 9 8 7 6 5 4 3 2 1

Contents

Introduction

This volume contains 14 papers on computational fluid dynamics written between 1967 and 1982, in particular papers on vortex methods and the projection method, as well as papers on the numerical solution of problems in kinetic theory, combustion theory, and gas dynamics. A great deal of practical experience and theoretical understanding has accumulated in these fields in recent years, and a systematic exposition of current knowledge is difficult to write and would be difficult to read. I believe that some of the ideas in this field are easier to learn if they are presented in the simpler garb that preceded the development of sophisticated implementations and of a general mathematical theory, and this is the motivation for publishing the present collection. These papers explain, among other topics, how one might set up a discrete approximation of the Navier–Stokes equations for an incompressible fluid, build a vortex method, or solve a Riemann problem. Some of these papers are by now quite old, especially by the standards of a rapidly changing field, and the reader should be aware of the existence of a large literature that is more up-to-date. In the next few paragraphs I would like to present a short summary of what the papers collected here contain and make some suggestions for further reading. These suggestions reflect my own interests and do not provide a complete bibliography of computational fluid mechanics or even of the topics covered in this book.

The general theme of these papers is the numerical solution of the equations of fluid mechanics in circumstances where the viscosity is small; the big problem that is looming beyond the specific applications is the problem of turbulence. It is generally understood that turbulence in fluids is dominated by the mechanics of vorticity, and many of the methods are based on vortex representations of the flow. A number of them employ random numbers in some form; the major motivation for studying random algorithms is the belief that they may lead to effective methods for sampling a turbulent flow field. The sequence of papers on vortex methods ends here with a paper that represents a possible starting point for speculations on the structure of turbulent flow. The belief that turbulence can be best understood in physical space, by considering the interactions between physical structures, rather than in wave-number space, where one considers interactions between Fourier modes, is not universally held. The reader interested in turbulence theory should find additional sources of information.

The first paper in the book presents the artificial compressibility method for finding steady-state solutions of the Navier–Stokes equations for an incompressible fluid, with applications to thermal convection. Related ideas have been introduced by Temam [T1] and Yanenko. A recent application with a bibliography can be found in [R2]. The main idea is to find a compressible system that has the same steady state as the given incompressible system, but has the property that its steady state can be reached with less labor. More generally, since the limit of infinite sound speed that leads from a compressible to an incompressible flow is well behaved [K2], one can try to approximate incompressible flow by an appropriate compressible flow even far from a steady state. These ideas are closely related to the penalty method, which is more natural in the context of finite element methods [B9].

Paper 2 presents the projection method for incompressible flow. The idea is that the equation of continuity can be viewed as a constraint, and one can solve the equations step-by-step by first ignoring the constraint and then projecting the result on the space of incompressible flows. Some subtlety is required to formulate the boundary conditions for the intermediate step. This construction was partially inspired by the existence theory for the Navier–Stokes equations presented in [F1] and has by now become standard (and indeed, in [C5] I showed that any consistent and stable approximation to the Navier–Stokes equations in pressure—velocity variables is essentially equivalent to the projection method). The correct formulation of the projection solves the problem of finding numerical boundary conditions for the pressure. Standard references in which this method is discussed include [C13], [G9], [P2], and [T1]. In the paper the projection was implemented by relaxation with a checkerboard pattern, which was a reasonable methodology for the time; however, faster Laplace solvers can be adapted for this purpose, see, e.g., [A4]. As a sidelight, I would like to mention that this paper contains a discussion of the relations between the DuFort–Frankel scheme, checkerboard relaxation and matrix condition (A). An interesting second-order variant of the projection method is presented in [B7]. Further discussion and finite element versions can be found, e.g., in [G9] and [T1].

Paper 3 contains a convergence proof for the projection method. Further work can be found, e.g., in [T1]. The analysis in the paper has two elements that are still of current interest: It shows that convergence and stability depend on the approximations for the gradient and for minus the divergence being at least approximately adjoint, and it contains a model of error growth which says basically that the error accumulates slowly as long as it is small, but if it ceases to be small all hell can break loose. A fixed-point theorem is needed to get convergence in a strong sense. This analysis shows that it is important to obtain error bounds rather than merely weak convergence results. Furthermore, there is a large recent literature in which it is claimed that very poor approximations can provide qualitative models of turbulence; the analysis in this paper suggests that such claims should be viewed with great caution. There is an interesting proof of the validity of a time-discretized projection in [E1].

Paper 4 is a numerical study of Boltzmann's equation by a method that can be viewed as a numerical generalization of Grad's thirteen moment approximation. I am not sure that this is a very good method of solution (but I do not know what is). The method is also an early example of a combined expansion–collocation method (as in the pseudo-spectral method), it provides some insight into the usefulness of some of the standard analytical approximation for Boltzmann's equation, and it provides an interesting contrast to the lack of convergence of the seemingly related Wiener–Hermite expansions of turbulence theory (see, e.g., [C6]).

Paper 5 is the original paper on the random vortex method. An incompressible flow can be approximated by a collection of vortex elements. The main ideas in this paper are: the use of vortices with finite core to improve convergence, the use of a random walk to approximate diffusion, and vorticity creation at boundaries to represent the no-slip boundary condition. For reviews of the applications of this method, see, e.g., [A6], [G10], [M3],and [M4]. The method has been greatly improved in the years since this paper appeared and has been the object of a great amount of outstanding theoretical analysis. The generalizations to three space dimensions and more accurate boundary conditions will be discussed below. I would like to provide here references to some of the major developments:

(i) A beautiful convergence theory has been developed; for inviscid flow without boundaries, see [A5], [B2], [B3], [B4], [B5], [B6], [C18], [G6], [H1], [M4], and [R1]; ([A5] contains a review). For viscous flow, see [G5], [L1], and [L2]. The theory has suggested better choices of smoothing than the one I used in this paper (see in particular [B6] and [P1]). Convergence results for related problems can be found in [R3].

(ii) The algorithm in the paper is $O(N^2)$, i.e., it takes $O(N^2)$ operations to sum the interactions of N vortex elements. It turns out that one can sum these interactions, to within machine accuracy, by using only $O(N)$ or $O(N \log N)$ operations, thus greatly increasing the power of the method (see e.g. [A2], [G8], and [K1]).

(iii) Elaborate numerical checks on the accuracy of the method can be found, e.g., in [C1], [C2], [C3], [P1], and [S4], and applications include [B11] and [C2], [C3] and [G2] as well as the papers mentioned below. Interesting mixed methods are presented in [S5] and [V1].

(iv) General discussions of the role of vortex dynamics in the study of turbulence and more generally in fluid mechanics can be found in [C6], [M1], and [M2].

Paper 6 (with P. Bernard) describes a calculation of the motion of a vortex sheet represented by a collection of vortex blobs (i.e. vortex elements smoothed by a finite cut-off). Later papers that expanded and developed this approach include [A1], [C17], and [K4]. These latter exceptionally fine calculations have also led to important theoretical developments [D1].

Papers 7 and 8 describe applications of the random choice method to problems in compressible gas dynamics and combustion. The random choice method is essentially an implementation of a construction introduced by Glimm [G4] in a more theoretical setting. In this method Riemann problems are solved at each point in space and are then sampled to provide the solution of the equations at the next level in time. It is clear now that I was too sanguine as to the range of applicability of this method. Different ways of using the Riemann problem, based on the work of Godunov and Van Leer, have turned out to be more practical in most problems (for reviews, see, e.g., [B1] and [W1]). Indeed, the limitations of the approach and some improvements on it have been pointed out in [C15]. There are, however, situations where the more general approach cannot be relied upon to produce the right type of waves and the random choice approach may still be useful (see, e.g., [C8], [C16] and [T2]). Furthermore, the random choice approach is a hyperbolic analogue of the random vortex method and is thus of theoretical interest. There exist many generalizations of the Riemann solutions presented in these papers (see, e.g., [B8], [C8] and [T2]).

Paper 9 introduces the vortex sheet method for solving the Prandtl boundary layer equations. The significance of this method is that it provides a tool for creating vorticity at boundaries with great accuracy. The creation algorithm in paper six has a low order of accuracy, as proved, e.g., in [C14], because the no-slip boundary condition is imposed only at the end of each step and not continuously. To apply that condition continuously one needs a reflection principle, which exists for the Prandtl equations but not for the Navier–Stokes equations. The suggestion is therefore that the sheet algorithm should be used near walls and be coupled to a different interior method. In this paper, the interior method is the random choice method, but the natural candidate for an interior method is the vortex method (see, e.g., paper 11 below or [C2]). A further discussion of the relation between vorticity creation and other ways of imposing boundary conditions can be found in [A3] and [P4].

Paper 10 presents an algorithm for moving flames and an application to turbulent combustion. The algorithm is an implementation of a Huygens principle based on the SLIC interface algorithm [N1] (which is slightly improved here). A much more general relation between wave propagation and front motion was discovered and implemented in [S2] and [O3]. Further applications of the SLIC algorithm can be found in [C9]. The application to a model flame propagation problem shows how a turbulent flame can propagate through the entrainment of unburned fluid into vortical structures, with the so called "laminar flame velocity" playing a small role, if any. Much more elaborate versions of this idea have been suggested in [O1] and [O2]. An alternative approach to the study of moving fronts can be found in [C4].

Paper 11 contains a generalization of the vortex sheet and vortex blob methods to three space dimensions, based on a representation by vortex segments. Several features of this particular implementation can be greatly improved upon: The integration in time should be more accurate; higher-order cut-offs can be used; fast summation can be applied; the diagnostics could be much better. On the other hand, even this implementation reveals the importance of vortex hairpins near walls. Later work along these lines can be found, e.g., in [F1], and further discussions of three-dimensional vortex methods are presented in [B1], [F3], [K3], and [L1].

Paper 12 contains a summary of several numerical methods useful for combustion modeling, in particular a generalization of the random walk procedure that can be used to describe the transport of chemical species and heat. Other work along these lines can be found, e.g., in [G1], [G2], and [P3] as well as in the next paper.

Paper 13 is the outcome of a happy collaboration with Profs. Ghoniem and Oppenheim, in which vortex methods, the Huygens principle discussed above, and an algorithm for creating specific volume at a flame are used to solve a two-dimensional combustion problem under a range of simplifying assumptions. However, the algorithms do take into account the exothermic effects of combustion and the mutual interaction of the flame and fluid mechanics. This methodology has been widely used since then, see, e.g., [G2] and [S3].

Paper 14 contains a numerical simulation of the process of vortex folding. It supplements [C7] where a similar problem was handled with the help of scaling transformations. The calculations in paper 14 are imperfect; later numerical experiments showed in particular that a finer time resolution is needed and that the determination of the ϵ-support of the vorticity, and thus of its Hausdorff dimension, is ambiguous. Studies of folding, of its connection with energy conservation and with the renormalization of vortex calculations, have been pursued elsewhere with methods that rely in an essential way on non-numerical considerations and are thus outside the scope of the present collection of papers; see, e.g., [C7], [C10], [C11], [C12], and [S5]. The paper presented here is qualitatively correct, and demonstrates the ability of numerical methods to discover and explain unexpected phenomena. The phenomenon of vortex folding and filamentation may well be the key to the understanding of fluid turbulence yet was not part of the search for such understanding before computation revealed its ubiquity and importance.

The convergence of vortex calculations in three space dimensions is discussed, e.g., in [A5], [M4], and [L4]. A further discussion of the motion of vortex filaments can be found, e.g., in [B2], [C12], and [K3] and the references therein.

It is a pleasure to thank Profs. P. Bernard, A. Ghoniem, and A. K. Oppenheim for their permission to reproduce our joint work.

Bibliography

[A1] C. Anderson, ''A vortex method for flow with slight density variations,'' *J. Comp. Phys.,* **61** (1985), 417–444.

[A2] C. Anderson, ''A method of local corrections for computing the velocity field due to a distribution of vortex blobs,'' *J. Comp. Phys.,* **62** (1986), 111–123.

[A3] C. Anderson, ''Vorticity boundary conditions and boundary vorticity generation for two-dimensional incompressible viscous flow,'' *J. Comp. Phys.,* **80** (1989), 72–97.

[A4] C. Anderson, ''Numerical implementation of the projection method,'' manuscript, UCLA Math Dept., (1989).

[A5] C. Anderson and C. Greengard, ''On vortex methods,'' *SIAM J. Num. Anal.,* **22** (1985), 413–440.

[A6] C. Anderson and C. Greengard (editors), *Vortex Methods,* Lecture Notes in Mathematics, Vol. 1360, (Springer, New York, 1988).

[B1] S. Ballard, ''Parametrization of viscosity in three-dimensional vortex methods,'' in *Numerical Methods in Fluid Mechanics,* Morton and Baines, editors (Clarendon, Oxford, 1986).

[B2] J. T. Beale, ''A convergent 3D vortex method with grid-free stretching,'' *Math. Comp.,* **46** (1986), 401–424.

[B3] J. T. Beale and A. Majda, ''Rates of convergence for viscous splitting of the Navier–Stokes equations,'' *Math. Comp.,* **37** (1981), 243–260.

[B4] J. T. Beale and A. Majda, ''Vortex methods, I: Convergence in three dimensions,'' *Math. Comp.,* **39** (1982), 1–27.

[B5] J. T. Beale and A. Majda, ''Vortex methods, II: Higher-order accuracy in two and three dimensions,'' *Math. Comp.,* **32** (1982), 29–52.

[B6] J. T. Beale and A. Majda, ''High-order accurate vortex methods with explicit velocity kernels,'' *J. Comp. Phys.,* **58** 91985), 188–208.

[B7] J. Bell, P. Colella, and H. Glaz, ''A second-order projection method for the incompressible Navier–Stokes equations,'' *J. Comp. Phys.,* (1989), in press.

[B8] M. Ben-Artzi and A. Birman, ''Applications of the generalized Riemann problem to 1D compressible flow with material interfaces,'' *J. Comp. Phys.,* **65** (1986), 170–178.

[B9] M. Bercovici and M. Engelman, ''A finite element method for the numerical solution of viscous incompressible flows,'' *J. Comp. Phys.,* **30** (1979), 181–201.

[B10] D. L. Book, *Finite Difference Techniques for Vectorized Fluid Dynamics* (Springer, New York, 1981).

[B11] T. Buttke, ''Numerical study of superfluid turbulence in the self-induction approximation,'' *J. Comp. Phys.,* **76** (1988), 301–326.

[C1] C. C. Chang, ''Random vortex methods for the Navier–Stokes equations,'' *J. Comp. Phys.,* **76** (1988), 281–300.

[C2] A. J. Cheer, ''Numerical study of incompressible slightly viscous flow past bluff boldies and airfoils,'' *SIAM J. Sc. Stat. Comp.,* **4** (1983), 685–705.

[C3] A. J. Cheer, "Unsteady separated flow behind an impulsively started circular cylinder in slightly viscous fluid," *J. Fluid Mech.*, **201** (1989), 485–505.

[C4] I. L. Chern, J. Glimm, O. McBryan, B. Plohr, and S. Yaniv, "Front tracking for gas dynamics," *J. Comp. Phys.*, **62** (1986), 83–110.

[C5] A. J. Chorin, "Numerical solution of incompressible flow problems," *Studies in Numerical Analysis*, **2** (1968), 64–71.

[C6] A. J. Chorin, *Lectures on Turbulence Theory* (Publish/Perish, Berkeley, 1975).

[C7] A. J. Chorin, "Estimates of intermittency, spectra and blow-up in fully developed turbulence," *Comm. Pure Appl. Math.*, **34** (1981), 853–866.

[C8] A. J. Chorin, "The instability of fronts in a porous medium," *Comm. Math. Phys.*, **91** (1983), 103–116.

[C9] A. J. Chorin, "Curvature and solidification," *J. Comp. Phys.*, **57** (1985), 472–489.

[C10] A. J. Chorin, "Turbulence and vortex stretching on a lattice," *Comm. Pure Appl. Math.*, **39** (special issue, 1986), S47–S65.

[C11] A. J. Chorin, "Scaling laws in the lattice vortex model of turbulence," *Comm. Math. Phys.*, **114** (1988), 167–176.

[C12] A. J. Chorin, "spectrum, dimension and polymer analogies in fluid turbulence," *Phys. Rev. Lett.*, **60** (1988), 1947–1949.

[C13] A. J. Chorin and J. Marsden, *A Mathematical Introduction to Fluid Mechanics* (Springer, New York, 1979).

[C14] A. J. Chorin, T. J. Hughes, M. McCracken, and J. Marsden, "Product formulas and numerical algorithms," *Comm. Pure Appl. Math.*, **31** (1978), 205–256.

[C15] P. Colella, "Glimm's method for gas dynamics," *SIAM J. Sc. Stat. Comp.*, **3** (1982), 76–88.

[C16] P. Colella, A. Majda, and V. Roytburd, "Theoretical and numerical structure for reacting shock waves," *SIAM J. Sc. Stat. Comp.*, **7** (1986), 1059–1080.

[C17] R. Clements and D. Maull, "The representation of sheets of vorticity by discrete vortices," *Prog. Aero. Sc.*, **16** (1975), 129–146.

[C18] G. H. Cottet, "Méthodes particulaires pour l'équation d'Euler dans le plan," these de 3e cycle, Université M & P. Curie, (Paris, 1982).

[D1] R. DiPerna and A. Majda, "Concentration and regularization for 2D incompressible flow," *Comm. Pure Appl. Math.*, **60** (1987), 301–345.

[D2] C. Dritschel, "Contour surgery," *J. Comp. Phys.*, **77** (1988), 240–266.

[E1] D. Ebin and J. Marsden, "Groups of diffeomorphisms and incompressible flow," *Ann. Math.*, **92** (1970), 102–152.

[E2] R. Esposito and M. Pulvirenti, "Three-dimensional stochastic vortex flows," preprint (1987).

[F1] D. Fishelov, "Vortex methods for slightly viscous three-dimensional flows," *SIAM J. Sc. Stat. Comp.*, in press (1989).

[F2] H. Fujita and T. Kato, "On the Navier-Stokes initial value problem," *Arch. Rat. Mech. Anal.*, **16** (1964), 269–315.

[G1] A. Ghoniem and A. K. Oppenheim, ''Numerical solution for the problem of flame propagation by the random element method,'' *AIAA Jour.*, **22** (1984), 1429–1435.

[G2] A. Ghoniem, G. Heiderinejad, and A. Krishnan, ''Numerical simulation of a thermally stratified shear layer using the vortex element method,'' *J. Comp. Phys.*, **79** (1988), 135–166.

[G3] A. Ghoniem and F. Sherman, ''Grid-free simulation of diffusion using grid-free random walk methods,'' *J. Comp. Phys.*, **61** (1985), 1–38.

[G4] J. Glimm, ''Solutions in the large for nonlinear hyperbolic systems of conservation laws,'' *Comm. Pure Appl. Math.*, **18** (1965), 697–712.

[G5] J. Goodman, ''The convergence of random vortex methods,'' *Comm. Pure Appl. Math.*, **40** (1987), 189–220.

[G6] C. Greengard, ''Convergence of the vortex filament method,'' *Math. Comp.*, **47** (1986), 387–398.

[G7] C. Greengard, ''The core spreading vortex method approximates the wrong equation,'' *J. Comp. Phys.*, **61** (1985), 345–348.

[G8] L. Greengard and V. Rokhlin, ''A fast algorithm for particle simulations,'' *J. Comp. Phys.*, **73** (1987), 325–348.

[G9] P. Gresho and R. Sani, ''On the pressure boundary conditions for the incompressible Navier-Stokes equations,'' report UCRL-96741, Lawrence Livermore National Laboratory (Jan. 1987).

[G10] K. Gustaffson and J. Sethian, ''Vortex flows,'' *SIAM Publications*, 1990.

[H1] O. Hald, ''Convergence of vortex methods, II,'' *SIAM J. Sc. Stat. Comp.*, **16** (1979), 726–755.

[K1] J. Katzenelson, ''Computational structure of the N body problem,'' *SIAM J. Sc. Stat. Comp.*, in press (1989).

[K2] S. Klainerman and A. Majda, ''Compressible and incompressible fluids,'' *Comm. Pure Appl. Math.*, **35** (1982), 629–653.

[K3] O. Knio and A. Ghoniem, ''Three-dimensional vortex methods,'' *J. Comp. Phys.*, in press.

[K4] R. Krasny, ''Desingularization of periodic vortex sheet roll-up,'' *J. Comp. Phys.*, **65** (1986), 292–313.

[L1] A. Leonard, ''Computing three-dimentional vortex flows with vortex filaments,'' *Ann. Rev. Fluid Mech.*, **17** (1985), 523–559.

[L2] D. G. Long, ''Convergence of the random vortex method in two dimensions,'' *J. Amer. Math. Soc.*, **1** (1988), 779–804.

[L3] D. G. Long, ''Convergence of random vortex methods in three dimensions,'' preprint, Princeton (1988).

[M1] A. Majda, ''The mathematical foundations of incompressible flow,'' Lecture Notes, Princeton (1986).

[M2] A. Majda, ''Vorticity and the mathematical theory of incompressible fluid flow,'' *Comm. Pure Appl. Math.* (special issue 1986), S187–S220.

[M3] A. Majda, ''Vortex dynamics: Numerical analysis, scientific computing and mathematical theory,'' *Proceedings Int. Cong. Ind. and Appl. Math.*, Paris 1987, J. McKenna and R. Temam, editors (SIAM Publications, 1988).

[M4] C. Marchioro and M. Pulvirenti, *Vortex Methods in Two-Dimensional Fluid Mechanics*, Lecture Notes in Physics, Vol. 203, (Springer, New York, 1984).

[N1] W. Noh and P. Woodward, "A simple line interface algorithm," in *Proceedings 5th Int. Conf. Numerical Methods Fluid Dynamics,* A. Van de Vooren and P. Zandbergern, editors (Springer, New York, 1976).

[O1] A. K. Oppenheim, "Mechanics of turbulent flow in combustors for premixed gases," *Recent Advances in Aerospace Propulsion* (Springer, New York, 1989).

[O2] A. K. Oppenheim, "Dynamic features of combustion," *Phil. Trans. Roy. Soc. London,* **A315** (1985), 471–508.

[O3] S. Osher and J. Sethian, "Fronts propagating with curvature-dependent speed: Algorithms based on Hamilton-Jacobi formulations," *J. Comp. Phys.,* **79** (1988), 12–49.

[P1] M. Perlman, "On the accuracy of vortex methods," *J. Comp. Phys.,* **59** (1985), 200–223.

[P2] R. Peyret and T. Taylor, *Computational Methods for Fluid Flow* (Springer, New York, 1983).

[P3] E. G. Puckett, "Convergence of a random particle method to solutions of the Kolmogorov equation," *Math. Comp.,* **52** (1989), 615–645.

[P4] E. G. Puckett, "A study of the vortex sheet method and its rate of convergence," *SIAM J. Sc. Stat. Comp.,* **10** (1989), 289–327.

[R1] P. Raviart, "An analysis of particle methods," CIME course, Como, Italy, 1983.

[R2] A. Rizzi, "Multi-cell vortices computed in large scale difference solution to the incompressible Euler equations," *J. Comp. Phys.,* **77** (1988), 207–220.

[R3] S. Roberts, "Accuracy of the random vortex method for a problem with non-smooth initial conditions," *J. Comp. Phys.,* **58** (1985), 29–43.

[R4] S. Roberts, "Convergence of a random walk method for Burgers' equation," *Math. Comp.,* **52** (1989), 647–673.

[S1] J. Sethian, "Turbulent combustion in open and closed vessels," *J. Comp. Phys.,* **54** (1984), 425–457.

[S2] J. Sethian, "Curvature and the evolution of fronts," *Comm. Math. Phys.,* **101** (1985), 487–498.

[S3] J. Sethian, "Vortex methods and turbulent combustion," *Lectures in Applied Mathematics,* **22** (1985), 245–259.

[S4] J. Sethian and A. Ghoniem, "A validation study of vortex methods," *J. Comp. Phys.,* **74** (1988), 283–317.

[S5] A. Shestakov, "A hybrid vortex-ADI solution for flows of low viscosity," *J. Comp. Phys.,* **74** (1979), 313–332.

[S6] E. Siggia, "Collapse and amplification of a vortex filament," *Phys. Fluids,* **28** (1985), 794–804.

[T1] R. Temam, *The Navier-Stokes Equations* (Elsevier, Amsterdam, 1984).

[T2] Z. H. Teng, A. J. Chorin and T. P. Liu, "Riemann problems for reacting gas," *SIAM J. Appl. Math.,* **42** (1982), 964–981.

[V1] J. J. W. van der Vegt, "A variationally optimized vortex tracing algorithm for three-dimensional flows around solid bodies," Ph.D. thesis, Delft (1988).

[W1] P. Woodward and P. Colella, "Numerical simulation of fluid flow with strong shocks," *J. Comp. Phys.,* **54** (1984), 115–173.

A Numerical Method for Solving Incompressible Viscous Flow Problems[1]

ALEXANDRE JOEL CHORIN

Courant Institute of Mathematical Sciences,

New York University, New York, New York 10012

ABSTRACT

A numerical method for solving incompressible viscous flow problems is introduced. This method uses the velocities and the pressure as variables, and is equally applicable to problems in two and three space dimensions. The principle of the method lies in the introduction of an artificial compressibility δ into the equations of motion, in such a way that the final results do not depend on δ. An application to thermal convection problems is presented.

INTRODUCTION

The equations of motion of an incompressible viscous fluid are

$$\partial_t u_i + u_j \partial_j u_i = -\frac{1}{\rho_0}\partial_i p + \nu \Delta u_i + F_i\,, \quad \Delta \equiv \sum_j \partial_j^2,$$

$$\partial_j u_j = 0,$$

where u_i are the velocity components, p is the pressure, F_i are the components of the external force per unit mass, ρ_0 is the density, ν is the kinematic viscosity, t is the time, and the indices i, j refer to the space coordinates x_i, x_j, $i, j = 1, 2, 3$.

Let d be some reference length, and U some reference velocity; we write

$$u_i' = \frac{u_i}{U}, \quad x_i' = \frac{x_i}{d}, \quad p' = \left(\frac{d}{\rho_0 \nu U}\right) p, \quad F_i' = \frac{\nu U}{d^2} F_i\,, \quad t' = \left(\frac{\nu}{d^2}\right) t$$

and drop the primes, obtaining the dimensionless equations

$$\partial_t u_i + R u_j \partial_j u_i = -\partial_i p + \Delta u_i + F_i\,, \tag{1a}$$

$$\partial_j u_j = 0, \tag{1b}$$

[1] This work was partially supported by AEC Contract No. AT(30-1)-1480.

Reprinted with permission from *Journal of Computational Physics,* Vol. 2, No. 1, 1967.

where $R = Ud/\nu$ is the Reynolds number. Our purpose is to present a finite difference method for solving (1a)–(1b) in a domain D in two or three space dimensions, with some appropriate conditions prescribed on the boundary of D.

The numerical solution of these equations presents major difficulties, due in part to the special role of the pressure in the equations, and in part to the large amount of computer time which such solution usually requires, making it necessary to devise finite-difference schemes which allow efficient computation. In two-dimensional problems the pressure can be eliminated from the equations using the stream function and the vorticity, thus avoiding one of the difficulties. If, however, a solution in three space dimensions is desired, one is thrown back upon the primary variables, the velocities and the pressure. In what follows a numerical procedure using these variables is presented; it is equally applicable to two and three dimensional problems, and is believed to be computationally advantageous even in the two dimensional case. In the present paper we shall concentrate on the search for steady solutions of the equations; a related method for time-dependent problems will be presented in a forthcoming paper.

Methods using the velocities and the pressure in two dimensional incompressible flow problems have previously been devised. For example, in [4], Harlow and Welch follow a procedure which appears quite natural—and may indeed in their problem be quite appropriate. It runs as follows: Taking the divergence of Eqs. (1a) one obtains for the pressure an equation of the form

$$\Delta p = Q \qquad \Delta \equiv \sum \partial_j^2 \tag{2}$$

where Q is a quadratic function of the velocities, and eventually, a function also of the external forces. Boundary conditions for (2) can be obtained from (1a) applied at the boundary. There remains however the task of insuring that (1b) is satisfied. This is done by starting the calculation with velocity fields satisfying (1b), making sure that (1b) is always satisfied at the boundary, and solving (2) at every step so that (1b) remains satisfied as time is advanced. An ingenious formulation of the finite difference form of Equation (2) reduces considerably the arithmetic labor necessary to solve it.

In our opinion the main shortcoming of this procedure lies in its treatment of the boundary conditions. In order to satisfy the boundary conditions for (2) derived from (1a) and to satisfy (1b) near the boundary, it is necessary, in the finite-difference formulation, to assign values to the velocity fields at virtual points outside the boundary, and this in a situation where no reflection principle is known to hold.

Were this procedure to be used only for the purpose of obtaining an asymptotic steady solution (which was not the purpose in [4]), it would have additional shortcomings. It would be computationally wasteful to solve (2) at every intermediate step, and moreover, in many problems, to obtain an initial solution satisfying (1b) would be a major problem by itself.

We shall now present a method for solving the system (1a)–(1b), which we believe to be free of these difficulties, and computationally more efficient. We shall not use equation (2).

THE METHOD OF ARTIFICIAL COMPRESSIBILITY

We introduce the auxiliary system of equations

$$\begin{aligned} \partial_t u_i + R\partial_j(u_i u_j) &= -\partial_i p + \Delta u_i + F_i\,, \\ \partial_t \rho + \partial_j u_j &= 0,\ p = \rho/\delta. \end{aligned} \tag{3}$$

An alternative form for the first of these equations is

$$\partial_t u_i + R u_j \partial_j u_i = -\partial_i p + \Delta u_i + F_i \tag{3'}$$

We shall call ρ the artificial density, δ the artificial compressibility, and $p = \rho/\delta$ the artificial equation of state. t is an auxiliary variable whose role is analogous to that of time in a compressible flow problem.

If, as the calculation progresses, the solution of (3) converges to a steady solution, i.e. one which does not depend on t, this solution is a steady solution of (1) and does not depend on δ. δ appears as a disposable parameter, analogous to a relaxation parameter. The system (3) is not a purely artificial construction, as can be seen by comparing it with the equations of motion of a compressible fluid with a small Mach number.

Equations (3) contain an artificial sound speed

$$c = 1/\delta^{1/2}$$

and relative to that speed the artificial Mach number M is

$$M = \frac{R}{c} \max_D \left(\sum u_i^2\right)^{1/2}$$

It is clearly necessary that $M < 1$.

It now remains to replace the system (3) or (3′) by a finite difference system, and

(a) show that the finite difference approximation to (3) is stable,

(b) demonstrate that the solution of the difference system does indeed tend to a steady limit,

(c) find a value of δ and of any other parameter in the finite difference system such that the steady limit is reached as fast as possible, and show that the resulting procedure is indeed efficient.

(d) show that the steady limit of the difference system does tend to a steady solution of (1) as the mesh width tends to zero.

The author has not been able to carry out this program analytically, forcing heavy reliance on the numerical evidence.

It is not indispensable that the solution of the differential system (3) tend to a steady limit, as long as the solution of the difference system does. It is however believed that the solution of (3) does tend to a steady limit, at least in the absence of external forces, under quite general conditions. This can be proved in the limiting case $R = 0$, for problems in which the velocities are prescribed at the boundary. By linearity it is sufficient to consider the case of zero velocities at the boundary. From (3) the following equality can be obtained

$$\frac{1}{2} \partial_t \int_D \left(\frac{1}{2} u_i u_i + \frac{\rho^2}{\delta} \right) dV = - \int_D \sum_{i,j} (\partial_i u_j)^2 \, dV.$$

The integrands on both sides are positive; hence the u_i tend to the limit $u_i = 0$, and p to a limit independent of t. From (3) one sees that this limit is independent of the x_i and therefore is a constant.

The Finite-Difference Approximation

The system (3) can be used with various difference schemes. In the one adopted here, after some experimentation, the inertia and pressure terms are differenced according to the leap-frog scheme, i.e., both time and space derivatives are replaced by central differences, and the viscous dissipation terms are differenced according to the Dufort–Frankel pattern, in which a second derivative such as

$$\partial_1{}^2 u$$

is replaced by

$$\frac{1}{\Delta x_1{}^2} (u_{i+1}^n + u_{i-1}^n - u_i^{n+1} - u_i^{n-1}) \qquad u_i^n \equiv u(i\Delta x_1, n\Delta t).$$

Δt, Δx_1 are, respectively, the "time"- and space-variable increments.

Equations (3) then become, in the two dimensional case, and in the absence of external forces

$$\begin{aligned} u_{1(i,j)}^{n+1} - u_{1(i,j)}^{n-1} = &- R \frac{\Delta t}{\Delta x_1} ((u_{1(i+1,j)}^n)^2 - (u_{1(i-1,j)}^n)^2) \\ &- R \frac{\Delta t}{\Delta x_2} (u_{1(i,j+1)}^n \, u_{2(i,j+1)}^n - u_{1(i,j-1)}^n \, u_{2(i,j-1)}^n) \\ &+ \frac{2\Delta t}{\Delta x_1{}^2} (u_{1(i+1,j)}^n + u_{1(i-1,j)}^n - u_{1(i,j)}^{n+1} - u_{1(i,j)}^{n-1}) \end{aligned}$$

$$
\begin{aligned}
&+ \frac{2\Delta t}{\Delta x_2{}^2} (u^n_{1(i,j+1)} + u^n_{1(i,j-1)} - u^{n+1}_{1(i,j)} - u^{n-1}_{1(i,j)}) \\
&- \frac{\Delta t}{\Delta x_1} \frac{1}{\delta} (\rho^n_{i,j+1} - \rho^n_{i,j-1}), \\
u^{n+1}_{2(i,j)} - u^{n-1}_{2(i,j)} &= - R \frac{\Delta t}{\Delta x_1} (u^n_{1(i+1,j)} u^n_{2(i+1,j)} - u^n_{1(i-1,j)} u^n_{2(i-1,j)}) \\
&- R \frac{\Delta t}{\Delta x_2} ((u^n_{2(i,j+1)})^2 - (u^n_{2(i,j-1)})^2) \\
&+ \frac{2\Delta t}{\Delta x_1{}^2} (u^n_{2(i+1,j)} + u^n_{2(i-1,j)} - u^{n+1}_{2(i,j)} - u^{n-1}_{2(i,j)}) \\
&+ \frac{2\Delta t}{\Delta x_2{}^2} (u^n_{2(i,j+1)} + u^n_{2(i,j-1)} - u^{n+1}_{2(i,j)} - u^{n-1}_{2(i,j)}) \\
&- \frac{\Delta t}{\Delta x_2} \frac{1}{\delta} (\rho^n_{i,j+1} - \rho^n_{i,j-1}), \\
\rho^{n+1}_{i,j} - \rho^{n-1}_{i,j} &= - \frac{\Delta t}{\Delta x_1} (u_{1(i+1,j)} - u_{1(i-1,j)}) - \frac{\Delta t}{\Delta x_2} (u_{2(i,j+1)} - u_{2(i,j-1)}),
\end{aligned}
\tag{4}
$$

with

$$\rho^n_{i,j} \equiv \rho(i\Delta x_1, j\Delta x_2, n\Delta t), u^n_{m(i,j)} \equiv u_m(i\Delta x_1, j\Delta x_2, n\Delta t).$$

Similar expressions are used in the three-dimensional case.

It is also necessary to approximate the equation

$$\partial_t \rho = -\partial_j u_j$$

at the boundary. Suppose the boundary is the line $x_2 = 0$, represented by $j = 1$ (see Fig. 1). A reasonable approximation is:

$$\rho^{n+1}_{i,1} - \rho^{n-1}_{i,1} = -2 \frac{\Delta t}{\Delta x_2} (u^n_{2(i,2)} - u^n_{2(i,1)}) - \frac{\Delta t}{\Delta x_1} (u^n_{1(i+1,1)} - u^n_{1(i-1,1)}). \tag{5}$$

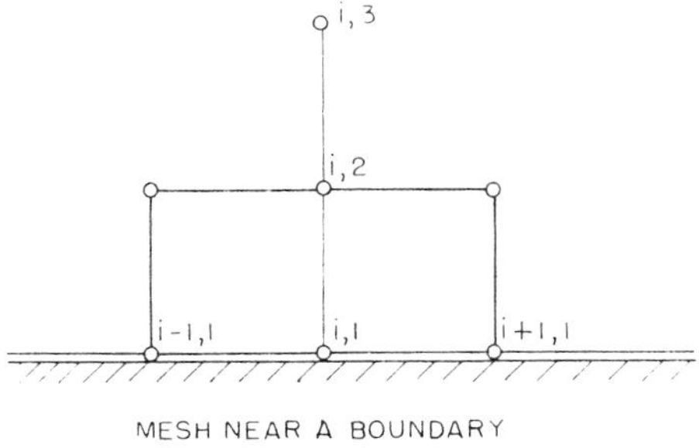

FIG. 1. Mesh near a boundary.

When $\rho^{n+1} \to \rho^{n-1}$, this expression tends to

$$2\,\frac{1}{\Delta x_2}\,(u^n_{2(i,2)} - u^n_{2(i,1)}) + \frac{1}{\Delta x_1}\,(u^n_{1(i+1,1)} - u^n_{1(i-1,1)}) = 0,$$

which approximates $\partial_j u_j = 0$ on the boundary to order Δx_2. A possible second-order approximation is

$$\rho^{n+1}_{i,1} - \rho^{n-1}_{i,1} = -4\,\frac{\Delta t}{\Delta x_2}\,(u^n_{2(i,2)} - u^{n+1}_{2(i,1)}) + \frac{\Delta t}{\Delta x_2}\,(u^n_{2(i,2)} - u^{n-1}_{2(i,1)}) - \frac{\Delta t}{\Delta x_1}\,(u^n_{1(i+1,1)} - u^n_{1(i-1,1)}).$$

One notices that these formulas contain three levels in appearance only, for, since $u^{n+1}_{i,j}$ does not depend on $u^n_{i,j}$, the calculation splits into two unrelated calculations on two intertwined meshes, one of which can be omitted. If this is done, the nth and $(n + 1)$st "time" levels can be considered as one level.

This scheme is stable for Δt small enough, and is entirely explicit. The presence of the dissipation terms suppresses the instabilities to which the nondissipative leap-frog scheme is susceptible. The known inaccuracy of the Dufort–Frankel scheme is of no relevance if only the asymptotic steady solution is sought. In fact, if we consider the Dufort–Frankel scheme

$$u^{n+1}_{i,j} - u^{n-1}_{i,j} = 2\,\frac{\Delta t}{\Delta x^2}\,(u^n_{i+1,j} + u^n_{i-1,j} + u^n_{i,j+1} + u^n_{i,j-1} - 2u^{n+1}_{i,j} - 2u^{n-1}_{i,j}) + 2\Delta t f, \tag{6}$$

which, for $\Delta t = o(\Delta x)$ approximates the equation

$$\partial_t u = \Delta u + f \qquad \Delta \equiv \partial_1{}^2 + \partial_2{}^2,$$

then, if we write

$$\omega = 8\,\frac{\Delta t}{\Delta x^2}\left(1 + 4\,\frac{\Delta t}{\Delta x^2}\right)^{-1}, \tag{7}$$

we see that (6) is nothing but the usual relaxation method for the solution of the 5-point Laplace difference equation, with relaxation parameter ω. Returning to the general system (4), we see that, since that system is stable only for Δt small enough, ω[defined by (7)] can take values only in an interval $0 \leqslant \omega \leqslant \omega_c < 2$. This is a familiar situation (see, e.g., [1]).

δ plays a role similar to that of a relaxation coefficient. Suppose the u_i are such that, at some point the finite-difference analog of (1b) is not satisfied; for example,

$$\frac{1}{\Delta x_1}\,(u^n_{1(i+1,j)} - u^n_{1(i-1,j)}) + \frac{1}{\Delta x_2}\,(u_{2(i,j+1)} - u_{2(i,j-1)}) < 0$$

so that $\rho_{i,j}^{n+1} > \rho_{i,j}^{n-1}$. Then a "density" gradient is formed which, through the terms $-\partial_i \rho/\delta$ in the momentum equations, will at the next step increase the velocity components pointing away from the point (i, j), thus increasing $\rho_{i,j}$ and bringing the equation of continuity closer to being satisfied. This sequence resembles a relaxation step.

A stability analysis of (4) shows that if the boundary conditions consist of prescribed velocities, the system is stable when

$$\max_D \max_{\phi_i} \left\{\frac{1}{2} \mid \Sigma \mid + \frac{1}{2} [\Sigma^2 + 4c^2(\psi_1{}^2 + \psi_2{}^2 + \psi_3{}^2)]^{1/2}\right\} \leqslant 1,$$

where

$$\psi_i = \frac{\Delta t}{\Delta x_i} \sin \phi_i \qquad 0 \leqslant \phi_i \leqslant 2\pi, \qquad i = 1, 2, 3,$$

$$\Sigma = u_1\psi_1 + u_2\psi_2 + u_3\psi_3 \qquad c = \delta^{-1/2}$$

If one ensures that the flow is subsonic with respect to the artificial sound speed, the above condition is satisfied when

$$\Delta t \leqslant \frac{2}{n^{1/2}(1 + 5^{1/2})} (\min_i \Delta x_i)\, \delta^{1/2},$$

where n is the number of space dimensions.

If other types of boundary conditions are imposed, e.g., if the derivatives of the velocities are prescribed at the boundary, one has to ensure that no instabilities arise due to boundary effects. For details, see [3].

In fact, we have at our disposal two parameters, Δt and δ, to be assigned values which make convergence to the steady solution as rapid as possible. The stability condition restricts the range of permissible values of these parameters.

Finally, the accuracy of the finite-difference scheme can be improved in two-dimensional problems with the use of staggered nets. This was not done here because our programs were written with three-dimensional problems in view. Slight modifications of the scheme were found necessary in problems involving singular points on the boundary.

A Simple Test Problem

The system (1) with $F_i = 0$ will now be solved in a square domain D: $0 \leqslant x_1 \leqslant 1$, $0 \leqslant x_2 \leqslant 1$ with the boundary conditions

$$u_1 = 4x_2(1 - x_2), \quad u_2 = 0 \quad \text{on the lines} \quad x_1 = 0 \quad \text{and} \quad x_1 = 1,$$

$$u_1 = u_2 = 0 \qquad\qquad \text{on the lines} \quad x_2 = 0 \quad \text{and} \quad x_2 = 1.$$

This is a simple problem, designed to test our method. The domain D represents a segment of a channel. The reference velocity in the Reynolds number is the maximum velocity in the channel, and the reference length d is the width of the channel. The steady solution is known analytically; it is

$$u_1 = 4x_2(1 - x_2), \quad u_2 = 0, \quad p = C - x_1 \quad \text{in} \quad D$$

where C is an arbitrary constant.

The equation of continuity is represented at the boundary by the formula (5). The higher-accuracy formula was also tried, and the results are very similar.

In Table I results of a sample computation are presented. The initial values for Eqs. (3)—or, if one prefers, the initial guess at the steady solution—are very

TABLE I: ERRORS IN TEST PROBLEM

N	$E(u_1)$	$E(u_2)$	$E(p)$
0	1.	0.	8.
100	0.1053	2.0×10^{-2}	7.04
200	1.03×10^{-2}	1.5×10^{-3}	0.61
300	7.7×10^{-4}	1.0×10^{-4}	0.22
400	7.1×10^{-5}	1.7×10^{-5}	1.6×10^{-2}
500	6.5×10^{-6}	3.8×10^{-6}	8.6×10^{-3}
600	1.2×10^{-6}	3.9×10^{-7}	7.8×10^{-3}
700	4.1×10^{-7}	1.2×10^{-7}	5.2×10^{-4}
800	7.2×10^{-8}	1.7×10^{-8}	1.3×10^{-4}
900	2.3×10^{-8}	6.5×10^{-9}	3.6×10^{-5}
1000	5.4×10^{-9}	1.5×10^{-9}	9.5×10^{-6}
1100	1.5×10^{-9}	4.2×10^{-10}	2.5×10^{-6}
1200	3.9×10^{-10}	1.1×10^{-10}	6.6×10^{-7}
1300	1.0×10^{-10}	$3. \times 10^{-11}$	1.7×10^{-7}
1400	$3. \times 10^{-11}$	$1. \times 10^{-11}$	4.6×10^{-8}
1500	$1. \times 10^{-11}$	less than 5×10^{-12}	1.2×10^{-8}

unfavorable: $u_1 = u_2 = 0$ everywhere except at the boundary, $p = 0$ everywhere. These initial values are very unfavorable because u_1 is discontinuous, and therefore, in the first steps, $\partial_t p$ becomes very large. Convergence is much faster when the initial values are smooth, or when they incorporate some advance knowledge regarding the final solution. These initial values were chosen to demonstrate the convergence of the procedure even under unfavorable conditions. In Table I the Reynolds number R is 1, $\delta = 0.00032$; 19 mesh points were used in each space direction. N is the number of steps, $E(u_1)$, $E(u_2)$, $E(p)$ are the errors, i.e., the maxima of the differences between the computed solution and the analytic solution given above. The constant C in the computed pressure is determined from the values of p on the line $x_1 = 0$.

It should be kept in mind that every step is very simple, being entirely explicit.

The optimal value of δ, δ_{opt}, has to be determined from a preliminary test computation; it is independent of Δx. Δt is determined from the relation

$$\Delta t = 0.6 \cdot \Delta x \cdot \delta^{1/2}$$

so that the stability requirement is met. δ_{opt} is not sharply defined; for $R = 0$, all values of δ between 0.006 and 0.05 lead to approximately the same rate of convergence.

The channel flow problem was solved for values of R varying between 0 and 1000. The method converged for all these values, although convergence was very slow for the higher values of R. δ_{opt} decreases as R increases.

The problem in this section is particularly simple; the analytic steady solution is known, and it satisfies the finite-difference equations exactly. The method was of course applied to less trivial problems, one of which will now be described.

Thermal Convection in a Fluid Layer Heated from Below. The Two-Dimensional Case

Suppose a plane layer of fluid, of thickness d and infinite lateral extent, in the field of gravity, is heated from below. The lower boundary $x_3 = 0$ is maintained at a temperature T_0, the upper boundary $x_3 = d$ at a temperature T_1, with $T_0 - T_1$ positive. (x_3 is the vertical coordinate.) The warmer fluid at the bottom of the layer expands, and tends to move upwards; this tendency is inhibited by the viscous stresses.

The equations governing the fluid motions are, in the Boussinesq approximation (see [2], [3]),

$$\partial_t u_i + u_j \partial_j u_i = -\frac{1}{\rho_0}\partial_i p + \nu \Delta u_i - g(1 - \alpha(T - T_0))\,\epsilon_i\,,$$
$$\partial_t T + u_j \partial_j T = k \Delta T, \qquad \partial_j u_j = 0,$$

where k is the coefficient of thermal conductivity, g is the force of gravitation, T the temperature, α the coefficient of thermal expansion of the fluid, and ϵ_i are the components of the unit vector pointing upwards.

We write

$$u_i' = \frac{d}{\nu}u_i\,, \qquad T' = \frac{T - T_1}{T_0 - T_1}\,, \qquad t' = \frac{\nu^2}{d}\,t,$$
$$x_i' = \frac{x_i}{d}\,, \qquad p' = \frac{1}{\rho_0}\left(\frac{d}{\nu}\right)^2 p$$

and drop the primes. The equations become

$$\partial_t u_i + u_j \partial_j u_i = -\partial_i p + \Delta u_i - \frac{R^*}{\sigma q}(1 - q(T-1))\,\epsilon_i\,,$$
$$\partial_t T + u_j \partial_j T = \frac{1}{\sigma}\Delta T, \qquad \partial_j u_j = 0 \tag{8}$$

where $R^* = [\alpha\beta g d^3(T_0 - T_1)](k\nu)^{-1}$ is the Rayleigh number, $\sigma = \nu/k$ the Prandtl number, and $q = \alpha(T_0 - T_1)$. It is assumed that the upper and lower boundaries are rigid, i.e., $u_i = 0$, $i = 1, 2, 3$ on $x_3 = 0$ and $x_3 = 1$.

It is known from the linearized stability theory that, for $R^* < R_c$, the state of rest is stable with respect to infinitesimal perturbations, where $R_c = 1707.62$ is the critical Rayleigh number (see [2]). This is taken to mean that for $R^* < R_c$ no convective motion can be maintained in the layer. When $R^* = R_c$ steady infinitesimal convection can first appear; the various field quantities are given by

$$u_3 = W(x_3)\,\phi,$$
$$u_1 = \frac{1}{a^2} W(x_3)\,\partial_1\phi, \qquad u_2 = \frac{1}{a^2} W(x_3)\,\partial_2\phi, \tag{9}$$
$$T = T(x_3)\,\phi,$$

where $\phi = \phi(x_1, x_2)$ determines the horizontal planform of the motion and satisfies

$$(\partial_1^2 + \partial_2^2)\,\phi = -a^2\phi.$$

$W(x_3)$, $T(x_3)$ are certain fully determined functions, $a = 3.117$, and the amplitude is of course undetermined.

In two dimensional convection $u_1 = 0$ and the motion is independent of x_1. ϕ has then the form

$$\phi = \cos a x_2$$

when $R^* = R_c$. The motion is periodic in x_2 with period $2\pi/a$. In this section we shall confine ourselves to two dimensional problems.

When $R^* > R_c$ it is known from experiment that steady convection sets in, at least when R^* is not too large. We shall assume that the motion remains periodic, with a period equal to the period of the first unstable mode (9) of the linearized theory. There is no difficulty in trying other periods. The periodicity assumption is physically very reasonable. We are interested in determining the amplitude of the motions, and more specifically, the magnitude of the heat transfer, measured by the dimensionless Nusselt number Nu. Nu is the ratio of the total heat transfer to the heat transfer which would have occurred if no convective motion were present; in our dimensionless variables it is simply

$$Nu = \frac{a}{2\pi}\int_0^{2\pi/a} (\sigma u_3 T - \partial_3 T)\,dx_2$$

and does not depend on x_3 when the convection is steady. For $R^* \leqslant R_c$, $Nu = 1$. It can be seen from (8) that the only physical parameters in the problem are R^* and σ; the solution does not depend on q, except inasmuch as R^* depends on q. Changing q in (8) simply implies a change in the definition of the pressure. We shall study the dependence of Nu on R^* and σ.

The auxiliary system used for finding steady solutions of (8) is

$$\begin{aligned} \partial_t u_i + u_j \partial_j u_i &= -\partial_i p + \Delta u_i - \frac{R^*}{\sigma q}(1 - q(T-1))\,\epsilon_i\,, \\ \partial_t T + u_j \partial_j T &= \frac{1}{\sigma}\Delta T, \qquad \partial_t \rho = -\partial_j u_j \end{aligned} \tag{10}$$

with the artificial equation of state, either

$$p = \frac{R^*}{\sigma q \delta}(\rho - 1 - q(T-1))$$

or

$$p = (R^*/\sigma q \delta)\,\rho.$$

The artificial sound speed c is in both cases

$$c = (R/\sigma q \delta)^{1/2}$$

δ is the artificial compressibility. The results are not affected by which equation of state is used. It should be noted that t in (10) does not represent real time.

The finite-difference scheme is a straightforward extension of the scheme presented before, i.e., a combined leap-frog and Dufort–Frankel scheme. It was found that the steady state is reached with less computing effort when the nonlinear terms are differenced in a nonconservative form, as in (3′). It was also observed that the computation proceeds with greatest efficiency when Δt is as large as possible, and hence, in view of the stability condition, when c is as small as possible. Since the artificial Mach number M has to be smaller than 1, q and δ were chosen in practice so as to have $M \sim 0.5 - 0.8$, thus allowing for possible velocity overshoots. A rough trial computation was usually made for every class of problems to determine the order of magnitude of M.

For every value of R^* and σ it is necessary to determine how many mesh points are needed to produce an accurate value of Nu. Serious errors may ensue when too few points are used. Every series of calculations was therefore performed at least twice, and the results accepted only if they had been approximately reproduced by two different calculations with differing meshes. As is to be expected, the number of points required increases with the Rayleigh number.

The initial data for the various problems consist of a zero-order solution on which a perturbation is imposed. The zero-order solution is

$$u_2 = u_3 = 0, \qquad T = 1 - x_3\,,$$

with ρ and p obtained by solving numerically the finite-difference equations in the absence of motion. The perturbation which produces the fastest convergence to the steady solution was found to be one in which the temperature alone is perturbed, by adding to it a multiple of the temperature field of the first unstable mode of the linearized theory.

The steady state is assumed to have been reached when two conditions are satisfied: (a) The Nusselt number evaluated at the lower boundary has varied by less than 0.2% over 100 steps, and (b) the Nusselt number evaluated at the lower boundary and the Nusselt number evaluated at midlayer differ by less than 0.2%.

Table II displays the variation of Nu with R^* for $\sigma = 1$ (see also Fig. 2). M is the number of mesh points in the x_2 direction, and N the number of mesh points in the x_3 direction. These results are in good agreement with some results obtained by G. Veronis and P. Schneck [5].

TABLE II: Nu AS A FUNCTION OF R^*/R_c

R^*/R_c	$M = 30, N = 26$	$M = 30, N = 28$
2	1.754	1.759
3	2.093	2.099
4	2.309	2.317
5	2.478	2.482
6	2.608	2.620
7	2.728	2.735
8	2.833	2.841
9	2.927	2.936
10	3.008	3.021
11	3.086	3.098
12	3.161	3.172
13	3.232	3.241

Table III gives an indication about the way Nu varies with σ, for $R^*/R_c = 7$. It is seen that Nu does not vary very much with σ, as already discovered by Veronis [6] with another type of boundary conditions.

For the sake of completeness, typical isotherm and stream line configurations are represented in Figs. 3(a) and (b). They were obtained with $R^*/R_c = 7$, $\sigma = 1$, $M = 30$, $N = 28$. The stream function ψ was obtained from the computed velocities, and affords a further check on the results, since the conditions

$$\partial_3\psi = 0, \qquad \partial_2\psi = 0$$

are satisfied at the upper and lower walls.

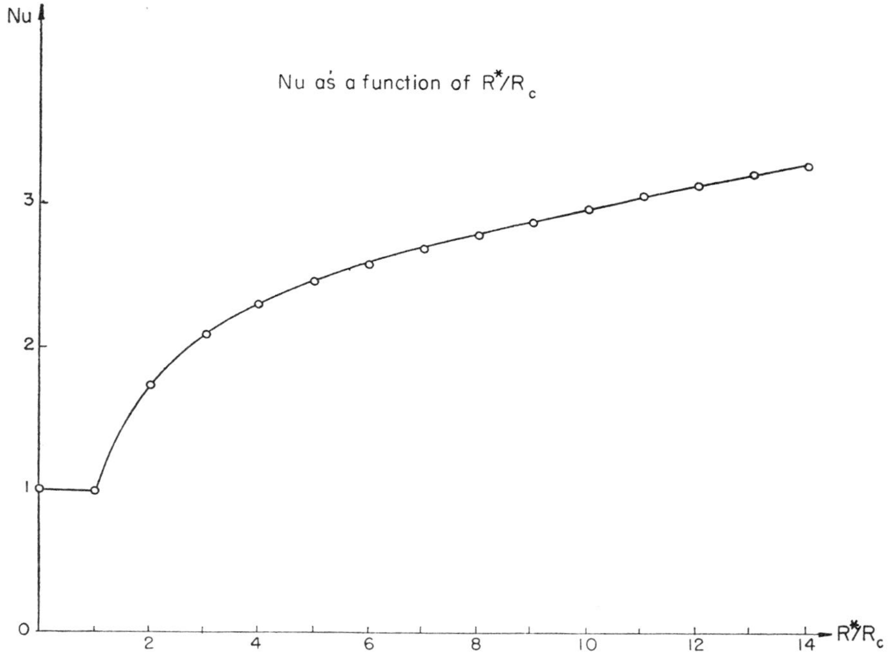

FIG. 2. *Nu* as a function of R^*/R_c.

TABLE III: *Nu* AS A FUNCTION OF σ

σ	$M = 30, N = 26$	$M = 30, N = 28$
20.0	2.67	2.68
6.8	2.68	2.69
1.0	2.73	2.73
0.2	2.68	2.68

THERMAL CONVECTION IN THREE SPACE DIMENSIONS.

In three space dimensions not only the amplitude of the motions is to be determined, but also their spatial configuration. For $R^* = R_c$ the function ϕ in (9) can be any periodic solution of

$$(\partial_1^2 + \partial_2^2 + a^2)\phi = 0, \qquad a = 3.117.$$

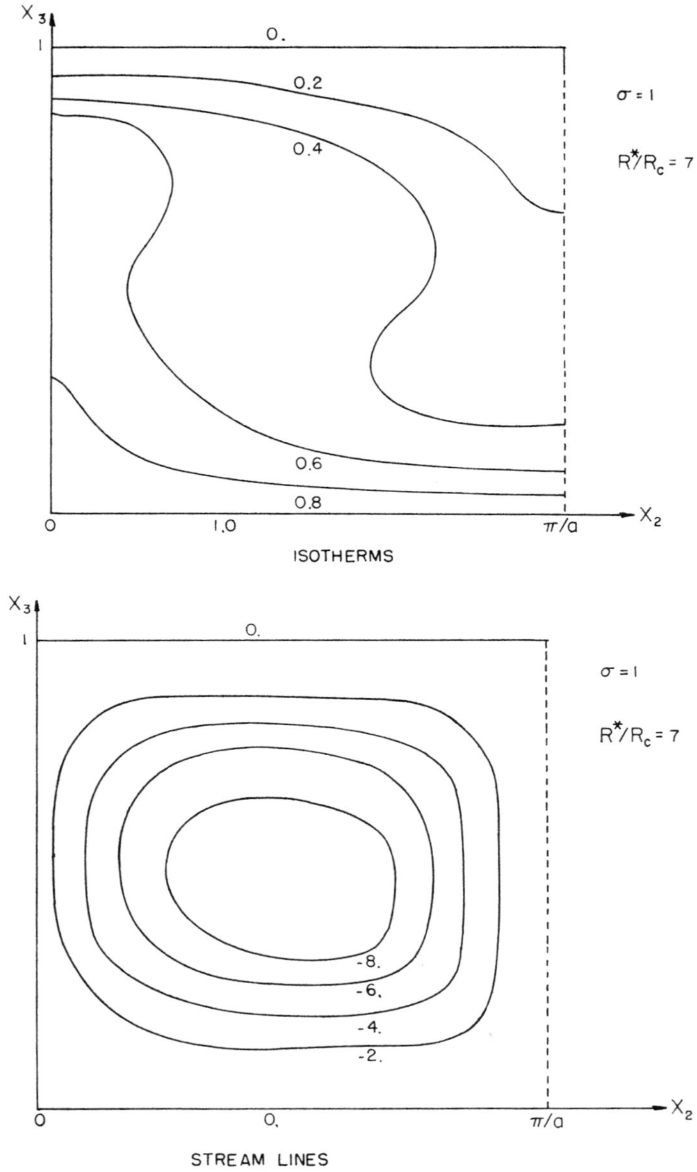

FIG. 3. (a) isotherms; (b) stream lines,

This corresponds to the fact that a given wave vector can be broken up into two orthogonal components in an infinite number of ways, with arbitrary amplitudes and phases. It is reasonable to assume that the cell patterns are made up of polygons whose union covers the (x_1, x_2)-plane; possible cell shapes are hexagons, rectangles, and rolls (i.e., two-dimensional convection cells). For $R^* > R_c$, the nonlinear terms in the equations determine which cell pattern actually occurs.

The numerical method described in this article is applicable; some computational results were described in [3]. The conclusion to be drawn from them is that the preferred cellular mode is the roll, but that even this preferred mode is subject to instabilities. A search for possible values of R, σ and a for which such instabilities do not occur will be described elsewhere.

Acknowledgments

The author wishes to thank Professor Peter D. Lax for his guidance during the execution of this work, which is based on results obtained in an NYU Ph. D. Thesis. The author also wishes to thank Professor Herbert B. Keller for several helpful discussions.

References

1. A. Brandt and J. Gillis, *Phys. Fluids* **9**, 690 (1966).
2. S. Chandrasekhar, "Hydrodynamic and Hydromagnetic Stability." The Clarendon Press, Oxford, England, 1961.
3. A. J. Chorin, AEC Research and Development Report No. NYO-1480-61, New York University (1966).
4. F. H. Harlow and J. E. Welch, *Phys. Fluids* **8**, 2182 (1965).
5. P. Schneck and G. Veronis, *Phys. Fluids* **10**, 927 (1967).
6. G. Veronis, *J. Fluid Mech.* **26**, 49 (1966).

Numerical Solution of the Navier-Stokes Equations*

By Alexandre Joel Chorin

Abstract. A finite-difference method for solving the time-dependent Navier-Stokes equations for an incompressible fluid is introduced. This method uses the primitive variables, i.e. the velocities and the pressure, and is equally applicable to problems in two and three space dimensions. Test problems are solved, and an application to a three-dimensional convection problem is presented.

Introduction. The equations of motion of an incompressible fluid are

$$\partial_t u_i + u_j \partial_j u_i = -\frac{1}{\rho_0}\partial_i p + \nu \nabla^2 u_i + E_i, \qquad \left(\nabla^2 \equiv \sum_j \partial_j^2\right), \qquad \partial_j u_j = 0,$$

where u_i are the velocity components, p is the pressure, ρ_0 is the density, E_i are the components of the external forces per unit mass, ν is the coefficient of kinematic viscosity, t is the time, and the indices i, j refer to the space coordinates x_i, x_j, $i, j = 1, 2, 3$. ∂_i denotes differentiation with respect to x_i, and ∂_t differentiation with respect to the time t. The summation convention is used in writing the equations.

We write

$$u_i' = \frac{u_i}{U}, \qquad x_i' = \frac{x_i}{d}, \qquad p' = \left(\frac{d}{\rho_0 \nu U}\right)p$$

$$E_i' = \left(\frac{\nu U}{d^2}\right)E_i, \qquad t' = \left(\frac{\nu}{d^2}\right)t,$$

where U is a reference velocity, and d a reference length. We then drop the primes. The equations become

$$\partial_t u_i + R u_j \partial_j u_i = -\partial_i p + \nabla^2 u_i + E_i, \tag{1}$$

$$\partial_j u_j = 0, \tag{2}$$

where $R = Ud/\nu$ is the Reynolds number. It is our purpose to present a finite-difference method for solving these equations in a bounded region $\mathfrak{D}$, in either two- or three-dimensional space. The distinguishing feature of this method lies in the use of Eqs. (1) and (2), rather than higher-order derived equations. This makes it possible to solve the equations and to satisfy the imposed boundary conditions while achieving adequate computational efficiency, even in problems involving three space variables and time. The author is not aware of any other method for which such claims can be made.

Received February 5, 1968.

* The work presented in this report is supported by the AEC Computing and Applied Mathematics Center, Courant Institute of Mathematical Sciences, New York University, under Contract AT(30-1)-1480 with the U. S. Atomic Energy Commission.

Reprinted from "Numerical Solution of the Navier-Stokes Equations," A.J.Chorin, *Mathematics of Computation*, (1968), Volume 22, Pages 742–762, by permission of the American Mathematical Society.

Principle of the Method. Equation (1) can be written in the form

$$\partial_t u_i + \partial_i p = \mathfrak{F}_i u \,, \tag{(1)'}$$

where $\mathfrak{F}_i u$ depends on u_i and E_i but not on p; Eq. (2) can be differentiated to yield

$$\partial_i(\partial_t u_i) = 0 \,. \tag{(2)'}$$

The proposed method can be summarized as follows: the time t is discretized; at every time step $\mathfrak{F}_i u$ is evaluated; it is then decomposed into the sum of a vector with zero divergence and a vector with zero curl. The component with zero divergence is $\partial_t u_i$, which can be used to obtain u_i at the next time level; the component with zero curl is $\partial_i p$. This decomposition exists and is uniquely determined whenever the initial value problem for the Navier-Stokes equations is well posed; it has also been extensively used in existence and uniqueness proofs for the solution of these equations (see e.g. [1]).

Let u_i, p denote not only the solution of (1) and (2) but also its discrete approximation, and let Du be a difference approximation to $\partial_j u_j$. It is assumed that at time $t = n\,\Delta t$ a velocity field u_i^n is given, satisfying $Du^n = 0$. The task at hand is to evaluate u_i^{n+1} from Eq. (1), so that $Du^{n+1} = 0$.

Let $Tu_i \equiv bu_i^{n+1} - Bu_i$ approximate $\partial_t u_i$, where b is a constant and Bu_i a suitable linear combination of u_i^{n-j}, $j \geqq 0$. An auxiliary field u_i^{aux} is first evaluated through

$$bu_i^{\text{aux}} - Bu_i = F_i u \tag{3}$$

where $F_i u$ approximates $\mathfrak{F}_i u$. u_i^{aux} differs from u_i^{n+1} because the pressure term and Eq. (2) have not been taken into account. u_i^{aux} may be evaluated by an implicit scheme, i.e. $F_i u$ may depend on u_i^n, u_i^{aux} and intermediate fields, say u_i^*, u_i^{**}. $bu_i^{\text{aux}} - Bu_i$ now approximates $\mathfrak{F}_i u$ to within an error which may depend on Δt.

Let $G_i p$ approximate $\partial_i p$. To obtain u_i^{n+1}, p^{n+1} it is necessary to perform the decomposition

$$F_i u = bu_i^{\text{aux}} - Bu_i = Tu_i + G_i p^{n+1} \,, \qquad D(Tu) = 0 \,.$$

It is, however, assumed that $Du^{n-j} = 0$, $j \geqq 0$. It is necessary therefore only to perform the decomposition

$$u_i^{\text{aux}} = u_i^{n+1} + b^{-1} G_i p^{n+1} \,, \tag{4}$$

where $Du_i^{n+1} = 0$, and u_i^{n+1} satisfies the prescribed boundary conditions. Since p^n is usually available and is a good first guess for the values of p^{n+1}, the decomposition (4) is probably best done by iteration. For that purpose we introduce the following iteration scheme:

$$u_i^{n+1,m+1} = u_i^{\text{aux}} - b^{-1} G_i^m p \,, \qquad m \geqq 1 \,, \tag{5a}$$

$$p^{n+1,m+1} = p^{n+1,m} - \lambda D u^{n+1,m+1} \,, \qquad m \geqq 1 \,, \tag{5b}$$

where λ is a parameter, $u_i^{n+1,m+1}$ and $p^{n+1,m+1}$ are successive approximations to u_i^{n+1}, p^{n+1}, and $G_i^m p$ is a function of $p^{n+1,m+1}$ and $p^{n+1,m}$ which converges to $G_i p^{n+1}$ as $|p^{n+1,m+1} - p^{n+1,m}|$ tends to zero. We set

$$p^{n+1,1} = p^n .$$

The iterations (5a) are to be performed in the interior of $\mathfrak{D}$, and the iterations (5b) in $\mathfrak{D}$ and on its boundary.

It is evident that (5a) tends to (4) if the iterations converge. We are using $G_i{}^m p$ instead of $G_i p$ in (5a) so as to be able to improve the rate of convergence of the iterations. This will be discussed in detail in a later section. The form of Eq. (5b) was suggested by experience with the artificial compressibility method [2] where, for the purpose of finding steady solutions of Eqs. (1) and (2), p was related to u_i by the equation

$$\partial_t p = \text{constant} \cdot (\partial_j u_j) .$$

When for some l and a small predetermined constant ϵ

$$\max_{\mathfrak{D}} |p^{n+1,l+1} - p^{n+1,l}| \leqq \epsilon$$

we set

$$u_i{}^{n+1} = u_i{}^{n+1,l+1} , \qquad p^{n+1} = p^{n+1,l+1} .$$

The iterations (5) ensure that Eq. (1), including the pressure term, is satisfied inside $\mathfrak{D}$, and Eq. (2) is satisfied in $\mathfrak{D}$ and on its boundary.

The question of stability and convergence for methods of this type has not been fully investigated. I conjecture that the over-all scheme which yields $u_i{}^{n+1}$ in terms of $u_i{}^n$ is stable if the scheme

$$T u_i = F_i u$$

is stable. The numerical evidence lends support to this conjecture.

We shall now introduce specific schemes for evaluating $u_i{}^{\text{aux}}$ and specific representations for Du, $G_i p$, $G_i{}^m p$. Many other schemes and representations can be found. The ones we shall be using are efficient, but suitable mainly for problems in which the boundary data are smooth and the domain $\mathfrak{D}$ has a relatively simple shape.

Evaluation of u_i^{aux}. We shall first present schemes for evaluating u_i^{aux}, defined by (3).

Equation (3) represents one step in time for the solution of the Burgers equation

$$\partial_t u_i = \mathfrak{F}_i u ,$$

which can be approximated in numerous ways. We have looked for schemes which are convenient to use, implicit, and accurate to $O(\Delta t) + O(\Delta x^2)$, where Δx is one of the space increments Δx_i, $i = 1, 2, 3$. Implicit schemes were sought because explicit ones typically require, in three space dimensions, that

$$\Delta t < \tfrac{1}{6} \Delta x^2$$

which is an unduly restrictive condition. On the other hand, implicit schemes of accuracy higher than $O(\Delta t)$ would require the solution of nonlinear equations at every step, and make it necessary to evaluate $u_i{}^{\text{aux}}$ and $u_i{}^{n+1}$ simultaneously rather than in succession. Since we assume throughout that $\Delta t = O(\Delta x^2)$, the gain in accuracy would not justify the effort.

Two schemes have been retained after some experimentation. For both of them

$$Tu_i \equiv (u_i^{n+1} - u_i^n)/\Delta t\,, \qquad (b^{-1} \equiv \Delta t\,, \quad Bu_i \equiv u_i^n/\Delta t)\,.$$

They are both variants of the alternating direction implicit method.

(A) In two-dimensional problems we use a Peaceman-Rachford scheme, as proposed by Wilkes in [3] in a different context. This takes the form

$$\begin{aligned} u^*_{i(q,r)} = u^n_{i(q,r)} &- R\frac{\Delta t}{4\Delta x_1} u^n_{1(q,r)}(u^*_{i(q+1,r)} - u^*_{i(q-1,r)}) \\ &- R\frac{\Delta t}{4\Delta x_2} u^n_{2(q,r)}(u^n_{i(q,r+1)} - u^n_{i(q,r-1)}) \\ &+ \frac{\Delta t}{2\Delta x_1^2}(u^*_{i(q+1,r)} + u^*_{i(q-1,r)} - 2u^*_{i(q,r)}) \\ &+ \frac{\Delta t}{2\Delta x_2^2}(u^n_{i(q,r+1)} + u^n_{i(q,r-1)} - 2u^n_{i(q,r)}) \\ &+ \frac{\Delta t}{2} E_i\,, \end{aligned} \tag{6a}$$

$$\begin{aligned} u^{\text{aux}}_{i(q,r)} = u^*_{i(q,r)} &- R\frac{\Delta t}{4\Delta x_1} u^*_{1(q,r)}(u^*_{i(q+1,r)} - u^*_{i(q-1,r)}) \\ &- R\frac{\Delta t}{4\Delta x_2} u^*_{2(q,r)}(u^{\text{aux}}_{i(q,r+1)} - u^{\text{aux}}_{i(q,r-1)}) \\ &+ \frac{\Delta t}{2\Delta x_1^2}(u^*_{i(q+1,r)} + u^*_{i(q-1,r)} - u^*_{i(q,r)}) \\ &+ \frac{\Delta t}{2\Delta x_2^2}(u^{\text{aux}}_{i(q,r+1)} + u^{\text{aux}}_{i(q,r-1)} - 2u^{\text{aux}}_{i(q,r)}) \\ &+ \frac{\Delta t}{2} E_i\,, \end{aligned} \tag{6b}$$

where u_i^* are auxiliary fields, and $u_{i(q,r)} \equiv u_i(q\Delta x_1, r\Delta x_2)$. As usual, the one-dimensional systems of algebraic equations can be solved by Gaussian elimination.

(B) In two-dimensional and three-dimensional problems we use a variant of the alternating direction method analyzed by Samarskii in [4]. This takes the form

$$\begin{aligned} u^*_{i(q,r,s)} = u^n_{i(q,r,s)} &- R\frac{\Delta t}{2\Delta x_1} u^n_{1(q,r,s)}(u^*_{i(q+1,r,s)} - u^*_{i(q-1,r,s)}) \\ &+ \frac{\Delta t}{\Delta x_1^2}(u^*_{i(q+1,r,s)} + u^*_{i(q-1,r,s)} - 2u^*_{i(q,r,s)})\,, \end{aligned}$$

$$\begin{aligned} u^{**}_{i(q,r,s)} = u^*_{i(q,r,s)} &- R\frac{\Delta t}{2\Delta x_2} u^*_{2(q,r,s)}(u^{**}_{i(q,r+1,s)} - u^{**}_{i(q,r-1,s)}) \\ &+ \frac{\Delta t}{\Delta x_2^2}(u^{**}_{i(q,r+1,s)} + u^{**}_{i(q,r-1,s)} - 2u^{**}_{i(q,r,s)}) \end{aligned}$$

$$u^{\text{aux}}_{i(q,r,s)} = u^{**}_{i(q,r,s)} - R\frac{\Delta t}{2\Delta x_3} u^{**}_{3(q,r,s)}(u^{\text{aux}}_{i(q,r,s+1)} - u^{\text{aux}}_{i(q,r,s-1)})$$

$$+ \frac{\Delta t}{\Delta x_3^2}(u^{\text{aux}}_{i(q,r,s+1)} + u^{\text{aux}}_{i(q,r,s-1)} - 2u^{\text{aux}}_{i(q,r,s)})$$

$$+ \Delta t E_{i(q,r,s)} ,$$

$$u_{i(q,r,s)} \equiv u_i(q\Delta x_1, r\Delta x_2, s\Delta x_3) ,$$

$$E_{i(q,r,s)} \equiv E_i(q\Delta x_1, r\Delta x_2, s\Delta x_3) .$$

u_i^*, u_i^{**} are auxiliary fields. These equations can be written in the symbolic form

$$\begin{aligned}(I - \Delta t Q_1)u_i^* &= u_i^n ,\\ (I - \Delta t Q_2)u_i^{**} &= u_i^* ,\\ (I - \Delta t Q_3)u_i^{\text{aux}} &= u_i^{**} + \Delta t E_i ,\end{aligned} \tag{7}$$

where I is the identity operator, and Q_l involves differentiations with respect to the variable x_l only.

It can be verified that when $R = 0$ scheme (6) is accurate to $O(\Delta t^2) + O(\Delta x^2)$. When $R \neq 0$ however, they are both accurate to the same order. Scheme (7) is stable in three-dimensional problems; the author does not know of a simple extension of scheme (6) to the three-dimensional case. Scheme (7) has two useful properties: It requires fewer arithmetic operations per time step than scheme (6), and because of the simple structure of the right-hand sides, the intermediate fields u_i^*, u_i^{**} do not have to be stored separately.

If either scheme is to be used in a problem in which the velocities u_i^{n+1} are prescribed at the boundary, values of u_i^*, u_i^{**}, u_i^{aux} at the boundary have to be provided in advance so that the several implicit operators can be inverted. Consider the case of the scheme (7). We have

$$\begin{aligned}u_i^{n+1} &= (I + \Delta t Q_1 + \Delta t Q_2 + \Delta t Q_3)u_i^n + \Delta t E_i - \Delta t G_i p^n + O(\Delta t^2) ,\\ u_i^* &= (I + \Delta t Q_1)u_i^n + O(\Delta t^2) ,\\ u_i^{**} &= (I + \Delta t Q_1 + \Delta t Q_2)u_i^n + O(\Delta t^2) ,\\ u_i^{\text{aux}} &= (I + \Delta t Q_1 + \Delta t Q_2 + \Delta t Q_3)u_i^n + \Delta t E_i + O(\Delta t^2) .\end{aligned}$$

From these relations it can be deduced that if we set at the boundary

$$\begin{aligned}u_i^* &= u_i^{n+1} - \Delta t Q_2 u_i^{n+1} - \Delta t Q_3 u_i^{n+1} - \Delta t E_i + \Delta t \overline{G}_i p ,\\ u_i^{**} &= u_i^{n+1} - \Delta t Q_3 u_i^{n+1} - \Delta t E_i + \Delta t \overline{G}_i p ,\\ u_i^{\text{aux}} &= u_i^{n+1} + \Delta t G_i p ,\end{aligned} \tag{8}$$

the scheme will remain accurate to $O(\Delta t)$. Here $\overline{G}_i$ does not have to be identical with G_i; all we need is

$$\overline{G}_i p^n = G_i p^n + O(\Delta t) .$$

The reason for introducing the new operator $\overline{G}_i$ is that at the boundary the normal component of G_i has to be approximated by one-sided differences, while this is not necessary in the interior of the domain $\mathfrak{D}$ where Eq. (4) is assumed to hold.

More accurate expressions for the auxiliary fields at the boundaries can be used, provided one is willing to invest the additional programming effort required to implement them on the computer. Appropriate expressions for u_i^*, u_i^{aux} at the boundary can be derived for use with the scheme (6).

It should be noted that for problems in which the viscosity is negligible, it is possible to devise explicit schemes accurate to $O(\Delta t^2) + O(\Delta x^2)$ and stable when $\Delta t = O(\Delta x)$. Such schemes will be discussed elsewhere.

The Dufort-Frankel Scheme and Successive Point Over-Relaxation. In order to explain our construction of D, G_i^m and our choice of λ for use in (5a) and (5b), we need a few facts concerning the Dufort-Frankel scheme for the heat equation and its relation to the relaxation method for solving the Laplace equation.

Consider the equation

$$-\nabla^2 u = f\,, \qquad (\nabla^2 \equiv \partial_1^2 + \partial_1^2) \tag{9}$$

in some nice domain $\mathfrak{D}$, say a rectangle. u is assumed known on the boundary of $\mathfrak{D}$. We approximate this equation by

$$-Lu = f\,, \tag{10}$$

where L is the usual five-point approximation to the Laplacian, and u and f are now m-component vectors. m is the number of internal nodes of the resulting difference scheme. For the sake of simplicity we assume that the mesh spacings in the x_1 and x_2 directions are equal, $\Delta x_1 = \Delta x_2 = \Delta x$; this implies no essential restriction. The operator $-L$ is represented by an $m \times m$ matrix A.

We write

$$A = A' - E - E'$$

where E, E' are respectively strictly upper and lower triangular matrices, and A' is diagonal. The convergent relaxation iteration scheme for solving (10) is defined by

$$(A' - \omega E)u^{n+1} = \{(1 - \omega)A' + \omega E'\}u^n + \omega f \tag{11}$$

(see e.g. Varga [5]). ω is the relaxation factor, $0 < \omega < 2$, and the u^n are the successive iterates. The evaluation of the optimal relaxation factor ω_{opt} depends on the fact that A satisfies "Young's condition (A)," i.e. that there exists a permutation matrix P such that

$$P^{-1}AP = \Lambda - N\,, \tag{12}$$

where Λ is diagonal, and N has the normal form

$$\begin{pmatrix} 0 & G \\ G' & 0 \end{pmatrix}$$

the zero submatrices being square. Under this condition, ω_{opt} can be readily determined.

The matrix A depends on the order in which the components of u^{n+1} are computed from u^n. Changing that order is equivalent to transforming A into $P^{-1}AP$, where P is a permutation matrix.

We now consider the solution of (14) to be the asymptotic steady solution of

$$\partial_\tau u = \nabla^2 u + f \tag{13}$$

and approximate the latter equation by the Dufort-Frankel scheme

$$u_{q,r}^{n+1} - u_{q,r}^{n-1} = \frac{2\Delta\tau}{\Delta x^2}\,(u_{q+1,r}^{n} + u_{q-1,r}^{n} + u_{q,r+1}^{n} + u_{q,r-1}^{n} - 2u_{q,r}^{n+1} - 2_{q,r}^{n-1}) + 2\Delta\tau f\,,$$

$$u_{q,r}^{n} \equiv u(q\Delta x_1,\, r\Delta x_2,\, n\Delta\tau)$$

which approximates (13) when $\Delta t = o(\Delta x)$. Grouping terms, we obtain

$$\begin{aligned}\left(1 + 4\,\frac{\Delta\tau}{\Delta x^2}\right)u_{q,r}^{n+1} - \left(1 - 4\,\frac{\Delta\tau}{\Delta x^2}\right)u_{q,r}^{n-1}\\ = 2\,\frac{\Delta\tau}{\Delta x^2}\,(u_{q+1,r}^{n} + u_{q-1,r}^{n} + u_{q,r+1}^{n} + u_{q,r-1}^{n}) + 2\Delta\tau f\,.\end{aligned} \tag{14}$$

Since $u_{q,r}^{n}$ does not appear in (14), the calculation separates into two independent calculations on intertwined meshes, one of which can be omitted. When this is done, we can write

$$U^{n+1} = \begin{pmatrix} u^{2n} \\ u^{2n+1} \end{pmatrix} \qquad (U^{n+1} \text{ has } m \text{ components})\,.$$

If we then write

$$\omega = \frac{8\Delta\tau/\Delta x^2}{1 + 4\Delta\tau/\Delta x^2} \tag{15}$$

we see that the iteration (14) reduces to an iteration of the form (11) where the new components of U^{n+1} are calculated in an order such that A has the normal form (12). The Dufort-Frankel scheme appears therefore to be a particular ordering of the over-relaxation method whose existence is equivalent to Young's condition (A).

The best value of $\Delta\tau$, $\Delta\tau_{\text{opt}}$, can be determined from ω_{opt} and relation (15). We find that $\Delta\tau_{\text{opt}} = O(\Delta x)$, therefore for $\Delta\tau = \Delta\tau_{\text{opt}}$ the Dufort-Frankel scheme approximates, not Eq. (13), but rather the equation

$$\partial_\tau u = \nabla^2 u - 2\left(\frac{\Delta\tau}{\Delta x}\right)^2 \partial_\tau^{\,2} u + f\,.$$

This is the equation which Garabedian in [6] used to estimate ω_{opt}. It can be used here to estimate $\Delta\tau_{\text{opt}}$. These remarks obviously generalize to problems where $\Delta x_1 \neq \Delta x_2$ or where there are more than two space variables.

The following remark will be of use: We could have approximated Eq. (13) by the usual explicit formula

$$u_{q,r}^{n+1} - u_{q,r}^{n} = \frac{\Delta\tau}{\Delta x^2}\,(u_{q+1,r}^{n} + u_{q-1,r}^{n} + u_{q,r+1}^{n} + u_{q,r-1}^{n} - 4u_{q,r}^{n}) + \Delta\tau f \tag{16}$$

and used this formula as an iteration procedure for solving (10). The resulting iteration converges only when $\Delta\tau/\Delta x^2 < 1/4$, and the convergence is very slow. The rapidly converging iteration procedure (14) can be obtained from (16) by splitting the term $u_{q,r}^{n}$ on the right-hand side into $\frac{1}{2}(u_{q,r}^{n+1} + u_{q,r}^{n-1})$.

Representation of D, G_i and G_i^m, and the Iteration Procedure for Determining u_i^{n+1}, p^{n+1}. For the sake of clarity we shall assume in this section that the domain $\mathfrak{D}$ is two-dimensional and rectangular, and that the velocities are prescribed at the boundary. Extension of the procedure to three-dimensional problems is immediate, and extension to problems with other types of boundary conditions often possible. Stress-free boundaries and periodicity conditions in particular offer no difficulty. Domains of more complicated shape can be treated with the help of appropriate interpolation procedures.

Our first task is to define D. Let $\mathcal{B}$ denote the boundary of $\mathfrak{D}$ and $\mathcal{C}$ the set of mesh nodes with a neighbor in $\mathcal{B}$. In $\mathfrak{D} - \mathcal{B}$ we approximate the equation of continuity by centered differences, i.e. we set

$$Du \equiv \frac{1}{2\Delta x_1}\left(u_{1(q+1,r)} - u_{1(q-1,r)}\right) + \frac{1}{2\Delta x_2}\left(u_{2(q,r+1)} - u_{2(q,r-1)}\right) = 0\,. \tag{17}$$

At the points of $\mathcal{B}$ we use second-order one-sided differences, so that Du is accurate to $O(\Delta x^2)$ everywhere. Consider the boundary line $x_2 = 0$, represented by $j = 1$ (Fig. 1). We have on that line

$$\begin{aligned} Du \equiv {} & \frac{2}{\Delta x_2}\left[u_{2(q,2)} - u_{2(q,1)} - \tfrac{1}{4}\left(u_{2(q,3)} - u_{2(q,1)}\right)\right] \\ & + \frac{1}{2\Delta x_1}\left(u_{1(q+1,1)} - u_{1(q-1,1)}\right) = 0 \end{aligned} \tag{18}$$

with similar expressions at the other boundaries. Equation (17) states that the total flow of fluid into a rectangle of sides $2\Delta x_1$, $2\Delta x_2$ is zero. Equation (18) does not have this elementary interpretation.

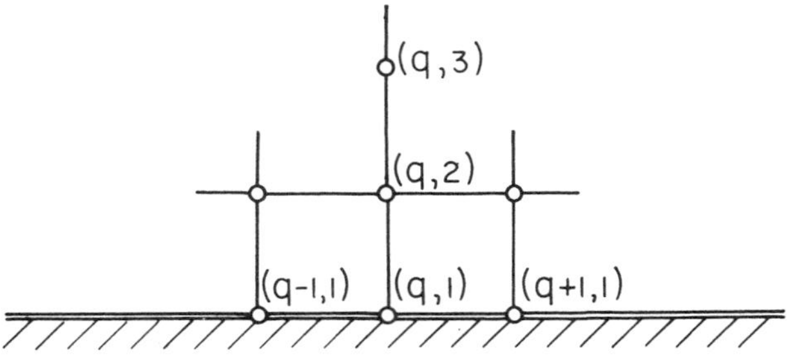

FIGURE 1. *Mesh Near a Boundary.*

We now define $G_i p$ at every point of $\mathfrak{D} - \mathcal{B}$ by

$$G_1 p = \frac{1}{2\Delta x_1}\left(p_{q+1,r} - p_{q-1,r}\right)\,,$$

$$G_2 p = \frac{1}{2\Delta x_2}\left(p_{q,r+1} - p_{q,r-1}\right)\,,$$

$$p_{q,r} \equiv p(q\Delta x_1, r\Delta x_2)\,,$$

i.e. $\partial_i p$ is approximated by centered differences. It should be emphasized that these forms of $G_i p$ and Du are not the only possible ones.

It is our purpose now to perform the decomposition (4). u_i^{n+1} is given on the boundary $\mathcal{B}$, u_i^{aux} is given in $\mathcal{D} - \mathcal{B}$ (the values of u_i^{aux} on $\mathcal{B}$, used in (6) or (7), are of no further use). p^{n+1} is to be found in $\mathcal{D}$ (including the boundary) and u^{n+1} in $\mathcal{D} - \mathcal{B}$, so that in $\mathcal{D} - \mathcal{B}$

$$u_i^{\text{aux}} = u_i^{n+1} + \Delta t G_i p$$

and in $\mathcal{D}$ (including the boundary)

$$Du^{n+1} = 0 .$$

This is to be done using the iterations (5), where the form of $G_i^m p$ has not yet been specified.

At a point (q, r) in $\mathcal{D} - \mathcal{B} - \mathcal{C}$, i.e. far from the boundary, one can substitute Eq. (5a) into Eq. (5b), and obtain

$$p^{n+1,m+1} - p^{n+1,m} = -\lambda Du^{\text{aux}} + \Delta t \lambda DG^m p . \tag{19}$$

This is an iterative procedure for solving the equation

$$Lp = \frac{1}{\Delta t} Du^{\text{aux}} , \tag{20}$$

where $Lp \equiv DGp$ approximates the Laplacian of p. With our choice of D and G_i, Lp is a five-point formula using a stencil whose nodes are separated by $2\Delta x_1$, $2\Delta x_2$. Equation (20) is of course a finite-difference analogue of the equation

$$\nabla^2 p = \partial_i \partial_j u_i u_j + \partial_j E_j , \tag{21}$$

which can be obtained from Eq. (1) by taking its divergence. At points of $\mathcal{B}$ or $\mathcal{C}$ if it is not possible to substitute (5a) into (5b) because at the boundary u_i^{n+1} is prescribed, $u^{n+1,m+1} = u^{n+1}$ for all m, (5a) does not hold and therefore (19) is not true. Near the boundary the iterations (5) provide boundary data for (20) and ensure that the constraint of incompressibility is satisfied. We proceed as follows: $G_i^m p$ and λ are chosen so that (19) is a rapidly converging iteration for solving (20); $G_i^m p$ at the boundary are then chosen so that the iterations (5) converge everywhere.

Let (q, r) again be a node in $\mathcal{D} - \mathcal{B} - \mathcal{C}$. $u_i^{n+1,m}$ and $p^{n+1,m}$ are assumed known. We shall evaluate simultaneously $p_{q,r}^{n+1,m+1}$ and the velocity components involved in the equation $Du^{n+1} = 0$ at (q, r), i.e. $u_{1(q\pm1,r)}^{n+1,m+1}$, $u_{2(q,r\pm1)}^{n+1,m+1}$ (Fig. 2). These velocity components depend on the value of p at (q, r) and on the values of p at other points. Following the spirit of the remark at the end of the last section, the value of p at (q, r) is taken to be

$$\tfrac{1}{2}(p_{q,r}^{n+1,m+1} + p_{q,r}^{n+1,m})$$

while at other points we use $p^{n+1,m}$.

This leads to the following formulae

$$p_{q,r}^{n+1,m+1} = p_{q,r}^{n+1,m} - \lambda Du^{n+1,m+1} \tag{22a}$$

(D given by (17))

$$u_{1(q+1,r)}^{n+1,m+1} = u_{1(q+1,r)}^{\text{aux}} - \frac{\Delta t}{2\Delta x_1}\left(p_{q+2,r}^{n+1,m} - \tfrac{1}{2}(p_{q,r}^{n+1,m+1} + p_{q,r}^{n+1,m})\right) , \tag{22b}$$

(22c) $$u_{1(q-1,r)}^{n+1,m+1} = u_{1(q-1,r)}^{\text{aux}} - \frac{\Delta t}{2\Delta x_1}\left(\tfrac{1}{2}(p_{q,r}^{n+1,m+1} + p_{q,r}^{n+1,m}) - p_{q-2,r}^{n+1,m}\right),$$

(22d) $$u_{2(q,r+1)}^{n+1,m+1} = u_{2(q,r+1)}^{\text{aux}} - \frac{\Delta t}{2\Delta x_2}\left(p_{q,r+2}^{n+1,m} - \tfrac{1}{2}(p_{q,r}^{n+1,m+1} + p_{q,r}^{n+1,m})\right),$$

(22e) $$u_{2(q,r-1)}^{n+1,m+1} = u_{2(q,r-1)}^{\text{aux}} - \frac{\Delta t}{2\Delta x_2}\left(\tfrac{1}{2}(p_{q,r}^{n+1,m+1} + p_{q,r}^{n+1,m}) - p_{q,r-2}^{n+1,m}\right).$$

These equations define $G_i^m p$. Clearly, $G_i^m p \to G_i p$.

Equations (22) can be solved for $p_{q,r}^{n+1,m+1}$, yielding

(23a) $$\begin{aligned} p_{q,r}^{n+1,m+1} = {} & (1 + \alpha_1 + \alpha_2)^{-1}[(1 - \alpha_1 - \alpha_2)p^{n+1,m} - \lambda D u^{\text{aux}} \\ & + \alpha_1(p_{q+2,r}^{n+1,m} + p_{q-2,r}^{n+1,m}) + \alpha_2(p_{q,r+1}^{n+1,m} + p_{q,r-2}^{n+1,m})], \end{aligned}$$

where $\alpha_i = \lambda \Delta t / 4 \Delta x_i$, $i = 1, 2$. This can be seen to be a Dufort-Frankel relaxation scheme for the solution of (20), as was to be expected. The $\Delta\tau$ of the preceding section is replaced here by $\lambda \Delta t/2$. Corresponding to $\Delta\tau_{\text{opt}}$ (or ω_{opt}) we find λ_{opt}. If p were known on $\mathcal{B}$ and $\mathcal{C}$, convergence of the iterations (23a) would follow from the discussion in the preceding section, and $\lambda = \lambda_{\text{opt}}$ would lead to the highest rate of convergence.

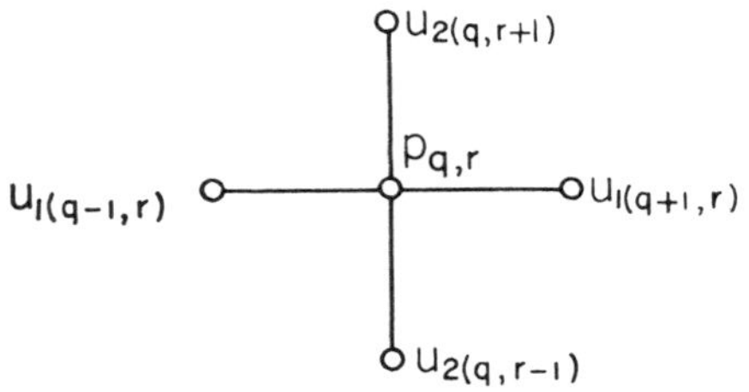

FIGURE 2. *Iteration Scheme*

In $\mathcal{B}$ and $\mathcal{C}$ formulae (22) are modified by the use of the known values of u_i^{n+1} at the boundary whenever necessary. This leads to the formulae, for (q, r) in $\mathcal{C}$:

(23b) $$\begin{aligned} p_{q,2}^{n+1,m+1} = {} & (1 + \alpha_1 + \tfrac{1}{2}\alpha_2)^{-1}[(1 - \alpha_1 - \tfrac{1}{2}\alpha_2)p_{q,2}^{n+1,m} - \lambda D u^{\text{aux}} \\ & + \alpha_1(p_{q+2,2}^{n+1,m} + p_{q-2,2}^{n+1,m}) + \alpha_2 p_{q,4}^{n+1,m}], \end{aligned}$$

and for (q, r) on $\mathcal{B}$:

(23c) $$\begin{aligned} p_{q,1}^{n+1,m+1} = {} & (1 + \alpha_1)^{-1}[(1 - \alpha_1)p_{q,1}^{n+1,m} - \lambda D u^{\text{aux}} \\ & + 2\alpha_2(p_{q,3}^{n+1,m} - \tfrac{1}{4}(p_{q,4}^{n+1,m} - p_{q,2}^{n+1,m}))] \end{aligned}$$

etc. In (23b) Du is given by (17), and in (23c) by (18). u_i^{aux} at the boundary is interpreted as u_i^{n+1}. Although no proof is offered, a heuristic argument and the numerical evidence lead us to state that the whole iteration system—Eqs. (23a), (23b), (23c)—converges for all $\lambda > 0$ and converges fastest when $\lambda \sim \lambda_{\text{opt}}$. None of the boundary instabilities which arise in two-dimensional vorticity-stream function calculation has been observed.

It can be seen that because our representation of $Du = 0$ expresses the balance of mass in a rectangle of sides $2\Delta x_i$, $i = 1, 2$, the pressure iterations split into two calculations on intertwined meshes, coupled at the boundary. The most efficient orderings for performing the iterations are such that the resulting over-all scheme is a Dufort-Frankel scheme for each one of the intertwined meshes. This involves no particular difficulty; a possible ordering for a rectangular grid is shown in Fig. 3. The iterations are to be performed until for some l

$$\max_{q,r} |p_{q,r}^{n+1,l+1} - p_{q,r}^{n+1,l}| \leqq \epsilon$$

for a predetermined ϵ.

The new velocities u_i^{n+1}, $i = 1, 2$, are to be evaluated using (22b), (22c), (22d), (22e). This has to be done only after the $p^{n+1,m}$ have converged. There is no need to evaluate and store the intermediate fields $u_i^{n+1,m+1}$. A saving in computing time can be made by evaluating Du^{aux} at the beginning of each iteration. We notice two advantages of our iteration procedure: Du^{n+1} can be made as small as one wishes independently of the error in Du^n; and when $p^{n+1,l+1}$ and $p^{n+1,l}$ differ by less than ϵ, $Du^{n+1} = O(\epsilon/\lambda)$; it can be seen that $\lambda_{\text{opt}} = O(\Delta x^{-1})$, hence $Du^{n+1} = O(\epsilon \Delta x)$. A gain in accuracy appears, which can be used to relax the convergence criterion for the iterations. This gain in accuracy is due to the fact that the u_i^{n+1} are evaluated using an appropriate combination of $p^{n+1,l}$ and $p^{n+1,l+1}$, rather than only the latest iterate $p^{n+1,l+1}$.

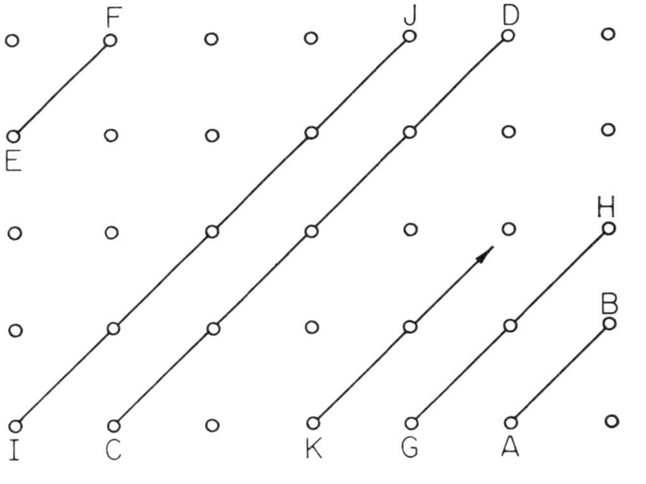

The domain is swept in the order AB, CD, EF, GH, IJ, K

FIGURE 3. *An Ordering for the Iteration Scheme*

Solution of a Simple Test Problem. The proposed method was first applied to a simple two-dimensional test problem, used as a test problem by Pearson in [7] for a vorticity-stream function method. $\mathfrak{D}$ is the square $0 \leqq x_i \leqq \pi$, $i = 1, 2$; $E_1 = E_2 = 0$; the boundary data are

$$u_1 = -\cos x_1 \sin x_2 e^{-2t}, \qquad u_2 = \sin x_1 \cos x_2 e^{-2t}.$$

The initial data are

$$u_1 = -\cos x_1 \sin x_2\,, \qquad u_2 = \sin x_1 \cos x_2\,.$$

The exact solution of the problem is

$$u_1 = -\cos x_1 \sin x_2 e^{-2t}\,, \qquad u_2 = \sin x_1 \cos x_2 e^{-2t}\,,$$
$$p = -R\,\tfrac{1}{4}(\cos 2x_1 + \cos 2x_2)e^{-4t}\,,$$

where R is the Reynolds number. This solution has the property

$$\partial_i p = -Ru_j \partial_j u_i\,;$$

hence curl (u_i) satisfies a linear equation. Nevertheless, this problem is a fair test of our method because $Du^{\text{aux}} \neq 0$.

We first evaluate λ_{opt}. For the equation

$$-Lu = f$$

in $\mathfrak{D}$, with a grid of mesh widths $2\Delta x_1$, $2\Delta x_2$, and u known on the boundary, we have

$$\omega_{\text{opt}} = \frac{2}{1 + (1 - \alpha^2)^{1/2}}\,,$$

where $\alpha = \frac{1}{2}(\cos 2\Delta x_1 + \cos 2\Delta x_2)$ is the largest eigenvalue of the associated Jacobi matrix (see [5]).

We put

$$q = \frac{\lambda_{\text{opt}}}{2}\left(\frac{\Delta t}{\Delta x_1^2} + \frac{\Delta t}{\Delta x_2^2}\right).$$

Equation (15) can be written as

$$\omega_{\text{opt}} = \frac{8q}{1 + 4q}\,,$$

therefore

$$q = \frac{1}{(1 - \alpha^2)^{1/2}}$$

and

$$\lambda_{\text{opt}} = \frac{4}{(\Delta t/\Delta x_1^2 + \Delta t/\Delta x_2^2)}\;\frac{1}{(1 - \alpha^2)^{1/2}}\,.$$

We now assume $\Delta x_1 = \Delta x_2 = \Delta x$, obtaining

$$\lambda_{\text{opt}} = \frac{2\Delta x^2}{\Delta t \sin\,(2\Delta x)}\,.$$

In Tables I, II, and III we display results of some sample calculations. n is the number of time steps; $e(u_i)$, $i = 1, 2$, are the maxima over $\mathfrak{D}$ of the differences between the exact and the computed solutions u_i. It is not clear how the error in the pressure is to be represented; p^n is defined at a time intermediate between $(n - 1)\,\Delta t$ and $n\,\Delta t$; it is proportional to R in our nondimensionalization. There are errors in p due to the fact, discussed at the end of the preceding section, that the iterations can be stopped before the $p^{n,m}$ have truly converged. $e(p)$ in the tables represents the maximum over the grid of the differences between the exact pressure at time $n\,\Delta t$ and the computed p^n, divided by R; it is given mainly for the sake of completeness. The accuracy of the scheme is to be judged by the smallness of $e(u_i)$. l is the number of iterations; it is to be noted that the first iteration always has to be performed in order that Eq. (1) be satisfied. "Scheme A" means that u_i^{aux} was evaluated using Eq. (6), and "Scheme B" means Eq. (7) were used.

TABLE I

Scheme A; $\Delta x = \pi/39$; $\Delta t = 2\,\Delta x^2 = 0.01397$; $\epsilon = \Delta x^2$; $R = 1$

n	$e(u_1)$	$e(u_2)$	$e(p)$	l
1	2.8×10^{-4}	2.6×10^{-4}	0.0243	1
2	2.7×10^{-4}	2.0×10^{-4}	0.0136	7
3	1.5×10^{-4}	1.3×10^{-4}	0.0069	4
4	1.8×10^{-4}	1.9×10^{-4}	0.0145	4
5	1.3×10^{-4}	1.7×10^{-4}	0.0089	5
6	1.3×10^{-4}	1.8×10^{-4}	0.0116	4
7	1.6×10^{-4}	1.9×10^{-4}	0.0144	4
9	1.4×10^{-4}	1.7×10^{-4}	0.0147	4
10	1.3×10^{-4}	1.6×10^{-4}	0.0156	4
20	1.8×10^{-4}	2.3×10^{-4}	0.0241	4

TABLE II

Scheme A; $\Delta x = \pi/39$; $\Delta t = 2\,\Delta x^2 = 0.01397$; $\epsilon = \Delta x^3$; $R = 1$

n	$e(u_1)$	$e(u_2)$	$e(p)$	l
1	8.5×10^{-5}	3.8×10^{-5}	0.0059	10
2	1.0×10^{-4}	5.7×10^{-5}	0.0067	10
3	1.0×10^{-4}	7.0×10^{-5}	0.0068	10
4	1.0×10^{-4}	7.8×10^{-5}	0.0068	10
5	1.0×10^{-4}	8.3×10^{-5}	0.0069	10
6	9.7×10^{-5}	8.6×10^{-5}	0.0070	10
7	9.4×10^{-5}	8.7×10^{-5}	0.0071	10
8	9.0×10^{-5}	8.7×10^{-5}	0.0073	10
9	8.7×10^{-5}	8.7×10^{-5}	0.0077	10
10	8.3×10^{-5}	8.5×10^{-5}	0.0082	10
20	1.0×10^{-4}	1.0×10^{-4}	0.0216	9

TABLE III

Scheme B; $\Delta x = \pi/39$; $\Delta t = \frac{1}{2}\Delta x^2 = 0.00324$; $\epsilon = \Delta x^2$; $R = 20$

n	$e(u_1)$	$e(u_2)$	$e(p)$	l
1	1.1×10^{-3}	1.2×10^{-3}	0.0217	15
3	1.9×10^{-3}	2.1×10^{-3}	0.0234	9
5	2.5×10^{-3}	2.8×10^{-3}	0.0242	9
7	3.3×10^{-3}	3.2×10^{-3}	0.0249	9
9	4.0×10^{-3}	3.5×10^{-3}	0.0253	8
20	5.8×10^{-3}	3.9×10^{-3}	0.0258	8

Tables I and II describe computations which differ only in the value of ϵ. They show that $\epsilon = \Delta x^2$ is an adequate convergence criterion. Table III indicates that fair results can be obtained even when $R\,\Delta t$ is fairly large; when $R = 20$, $\Delta x = \pi/39$, $\Delta t = \frac{1}{2}\Delta x^2$, we have

$$R \simeq 1.5\ \Delta x^{-1}\,.$$

The errors are of the order of 1%. Additional computational results were presented in [8].

Application to Thermal Convection. Suppose a plane layer of fluid, in the field of gravity, of thickness d and infinite lateral extent, is heated from below. The lower boundary $x_3 = 0$ is maintained at a temperature T_0, the upper boundary $x_3 = d$ at a temperature $T_1 < T_0$. The warmer fluid at the bottom expands and tends to move upward; this motion is inhibited by the viscous stresses.

In the Boussinesq approximation (see e.g. [9]) the equations describing the possible motions are

$$\partial_t u_i + u_j \partial_j u_i = -\frac{1}{\rho_0}\partial_i p + \nu \nabla^2 u_i - g(1 - (T - T_0))\delta_i\,,$$

$$\partial_t T + u_j \partial_j T = k\nabla^2 T\,, \qquad \partial_j u_j = 0\,,$$

where T is the temperature, k the coefficient of thermal conductivity, α the coefficient of thermal expansion, and δ_i the components of the unit vector pointing upwards.

We write

$$u_i' = \left(\frac{d}{\nu}\right)u_i\,, \qquad T' = \frac{T - T_1}{T_0 - T_1}\,, \qquad t' = \left(\frac{\nu^2}{d}\right)t\,,$$

$$x_i' = \frac{x_i}{d}\,, \qquad p' = \frac{1}{\rho_0}\left(\frac{d}{\nu}\right)^2 p + \frac{(T_1 - T_0)dgx_3}{\nu^2}$$

and drop the primes. The equations now are

$$\partial_t u_i + u_j \partial_j u_i = -\partial_i p + \nabla^2 u_i + \frac{R^*}{\sigma}(T - 1)\delta_i\,,$$

$$\partial_t T + u_j \partial_j T = \frac{1}{\sigma}\nabla^2 T\,, \qquad \partial_j u_j = 0\,,$$

where $R^* = \alpha g d^3(T_0 - T_1)k\nu$ is the Rayleigh number, and $\sigma = \nu/k$ the Prandtl number. The rigid boundaries are now situated at $x_3 = 0$ and $x_3 = 1$, where it is assumed that $u_i = 0$, $i = 1, 2, 3$.

It is known that for $R^* < R_c^*$, the state of rest is stable and no steady convection can arise, where $R_c^* = 1707.762$.

When $R^* = R_c^*$, steady infinitesimal convection can first appear, and the field quantities are given by

$$u_3 = CW(x_3)\phi\,,$$

$$u_i = \frac{C}{a^2}(\partial_3 W(x_3))\partial_i\phi\,, \qquad i = 1, 2\,,$$

$$T = CT(x_3)\phi$$

where $\phi = \phi(x_1, x_2)$ determines the horizontal planform of the motion and satisfies

$$(\partial_1^2 + \partial_2^2 + a^2)\phi = 0\,,$$

$W(x_3)$, $T(x_3)$ are fully determined functions of x_3, $a = 3.117$, and C is a small but undetermined amplitude.

In two-dimensional motion $u_1 = 0$ and the motion does not depend on x_1. We then have

$$\phi = \cos ax_2\,.$$

The motion is periodic in x_2 with period $2\pi/a$.

The Nusselt number Nu is defined as the ratio of the total heat transfer to the heat transfer which would have occurred if no convection were present. For $R^* \leqq R_c^*$, $Nu = 1$. In our dimensionless variables

$$Nu = \frac{a}{2\pi}\int_0^{2\pi/a} (u_3 T - \partial_3 T)dx_2\,.$$

A similar expression holds in the three-dimensional case. When the convection is steady Nu does not depend on x_3.

When $R^* > R_c^*$ steady cellular convection sets in. It is of interest to determine its magnitude and its spatial configuration. The problem of its magnitude, and in particular the dependence of Nu on R^* and σ, when the motion is steady, has been studied by the author in previous work [2], [10]. As to the shape of the convection cells, it is known that flows may exist in which the cells, when viewed from above, look like hexagons, or like rectangles with various ratios of length to width, or like rolls, i.e. two-dimensional convection cells (see [11]). However, only cellular structures which are stable with respect to small perturbations can persist in nature or be exhibited by our method. It has been shown, numerically by the author [10], experimentally by Koschmieder [12] and Rossby [13], theoretically, in the case of infinite σ and small perturbations, by Busse [14], that for $R^*/R_c^* < 10$ the preferred cellular mode is a roll. Busse showed that the rolls are stable for wave numbers in a certain range. We shall now demonstrate numerically the impermanence of hexagonal convection and the emergence of a roll.

Consider the case $R^*/R_c^* = 2$, $\sigma = 1$. We assume the motion to be periodic in the x_1 and x_2 directions, with periods respectively $4\pi/a\sqrt{3}$ and $4\pi/a$ (the first period is apparently in the range of stable periods for rolls as predicted by Busse). These are the periods of the hexagonal cells which could arise when $R^* = R_c^*$. The state of rest is perturbed by adding to the temperature in the plane $x_3 = \Delta x_3$ a multiple of the function $\phi(x_1, x_2)$ which corresponds to a hexagonal cell, and adding a small constant to the temperature on the line $x_1 = (3/4)(4\pi/a\sqrt{3})$, $x_2 = (3/4)(4\pi/a)$. We then follow the evolution of the convection in time, using a net of $24 \times 24 \times 25$, i.e.

$$\Delta x_1 = (4\pi/a\sqrt{3})/24\,, \qquad \Delta x_2 = (4\pi/a)/24\,, \qquad \Delta x_3 = 1/24\,.$$

We choose $\epsilon = \Delta x_2^2$, $\Delta t = 3\,\Delta x_3^2$. The convection pattern is visualized as follows: the velocities in the plane $x_3 = 17\,\Delta x_3$ are examined. If $u_{3(q,r,18)} > 0$ an $*$ is printed, if $u_{3(q,r,18)} \leqq 0$, a 0 is printed.

FIGURE 4. *Evolution of a Convection Cell*

```
*******000000000000000**
*******00000000000000000
*******0000000000000****
******00000000000000*0**
****00000000000000000***
000000000000000000000000
000000000000000000000000
000000000000000000000000
0000000000*0*00000000000
000000000000000000000000
000000*****0*00000000000
000000000000000000000000
000000000000000000000000
000000000000000000000000
000000000000000000000000
000000000000000000000000
0000000000000000***00000
0000000000000000***00000
0000000000000000***00000
000000000000000000000000
000000000000000000000000
000000000000000000000000
000000000000000000000000
000000000000000000000000
```

4a. *After* 1 *step* ($Nu = 1$)

```
*********0000000000000**
**********000000000*****
**********0000000*******
********000000000*******
******000000000000******
***000000000000000000***
000000000000000000000000
000000000*******00000000
00000************0000000
0000***********000000000
000**********00000000000
000********0000000000000
0000*****000000000000000
000000000000000000000000
000000000000000000000000
00000000000000*******000
0000000000000*********00
0000000000000*********00
0000000000000*********00
00000000000000*******000
000000000000000000000000
00000*000000000000000000
00*****00000000000000000
*******00000000000000000
```

4b. *After* 10 *steps* ($Nu = 1$)

```
**********00000000000000
***********000000000000*
***********000000000000*
***********000000000****
************************
***0000000**************
************************
**************0000000***
************000000000000
***********0000000000000
0**********0000000000000
00********00000000000000
00000***0000000000000000
0000000000000*******0000
00000000000***********00
00000000000************0
0000000000**************
0000000000**************
0000000000**************
00000000000*************
000000000000**********00
000000000000000000000000
00******0000000000000000
0*********00000000000000
```

4c. *After* 125 *steps* ($Nu = 1.25$)

```
**********00000000000000
***********000000000000*
************0000000000**
*************000000000**
**************00000000**
**************00000000**
**************00000000**
*************0000000000*
*************00000000000
00**********000000000000
0000********000000000000
0000000*****000000000000
00000000*******000000000
00000000**********000000
00000000*************000
00000000**************00
00000000***************0
00000000***************0
00000000***************0
000000000**************0
00000000000***********00
000000000000000000000000
000000000000000000000000
00*****00000000000000000
```

4d. *After* 225 *steps* ($Nu = 1.72$)

```
0000********000000000000
00*************000000000
0**************000000000
0***************00000000
0***************00000000
0***************00000000
00**************00000000
00*************000000000
000************000000000
0000***********000000000
00000**********000000000
00000**********000000000
00000***********00000000
00000************0000000
00000*************000000
00000*************000000
0000***************00000
00000**************00000
00000**************00000
000000************000000
0000000**********0000000
000000000******000000000
000000000000000000000000
000000000000000000000000
```

4e. *After* 325 *steps* ($Nu = 1.76$)

```
0000***********000000000
0000***********000000000
0000***********000000000
0000************00000000
000*************00000000
000*************00000000
000*************00000000
000*************00000000
0000***********000000000
0000***********000000000
0000***********000000000
0000***********000000000
0000***********000000000
0000***********000000000
000************000000000
000*************00000000
000*************00000000
000*************00000000
0000***********000000000
0000***********000000000
0000***********000000000
0000***********000000000
00000*********0000000000
00000*********0000000000
```

4f. *After* 430 *steps* ($Nu = 1.77$)

The evolution of the convection is shown in Figs. 4a, 4b, 4c, 4d, 4e, and 4f. The hexagonal pattern introduced into the cell is not preserved. The system evolves through various stages, and finally settles as a roll with period $4\pi/a\sqrt{3}$. The value of Nu evaluated at the lower boundary is printed at the bottom of each figure. The steady state value for a roll is 1.76. The final configuration of the system is independent of the initial perturbation. The calculation was not pursued until a completely steady state had been achieved because that would have been excessively time consuming on the computer. It is known from previous work that steady rolls can be achieved, and that the mesh used here provides an adequate representation.

Conclusion and Applications. The Benard convection problem is not considered to be an easy problem to solve numerically even in the two-dimensional case. The fact that with our method reliable time-dependent results can be obtained even in three space dimensions indicates that the Navier-Stokes equations do indeed lend themselves to numerical solution. A number of applications to convection problems, with or without rotation, can be contemplated; in particular, it appears to be of interest to study systematically the stability of Benard convection cells when $\sigma \neq \infty$, and when the perturbations have a finite amplitude.

Other applications should include the study of the finite amplitude instability of Poiseuille flow, the stability of Couette flow, and similar problems.

Acknowledgements. The author would like to thank Professors Peter D. Lax and Herbert B. Keller for their interest and for helpful discussions and comments.

New York University
Courant Institute of Mathematical Sciences
New York, New York 10012

1. H. Fujita & T. Kato, "On the Navier-Stokes initial value problem. I," *Arch. Rational Mech. Anal.*, v. 16, 1964, pp. 269–315. MR **29** #3774.

2. A. J. Chorin, "A numerical method for solving incompressible viscous flow problems," *J. Computational Physics*, v. 2, 1967, p. 12.

3. J. O. Wilkes, "The finite difference computation of natural convection in an enclosed cavity," Ph.D. Thesis, Univ. of Michigan, Ann Arbor, Mich., 1963.

4. A. A. Samarskiĭ, "An efficient difference method for solving a multi-dimensional parabolic equation in an arbitrary domain," *Ž. Vyčisl. Mat. i Mat. Fiz.*, v. 2, 1962, pp. 787–811 = *U.S.S.R. Comput. Math. and Math. Phys.*, v. 1963, 1964, no. 5, pp. 894–926. MR **32** #609.

5. R. Varga, *Matrix Iterative Analysis*, Prentice-Hall, Englewood Cliffs, N. J., 1962.

6. P. R. Garabedian, "Estimation of the relaxation factor for small mesh size," *Math. Comp.*, v. 10, 1956, pp. 183–185. MR **19**, 583.

7. C. E. Pearson, "A computational method for time dependent two dimensional incompressible viscous flow problems," Report No. SRRC-RR-64-17, Sperry Rand Research Center, Sudbury, Mass., 1964.

8. A. J. Chorin, "The numerical solution of the Navier-Stokes equations for incompressible fluid," AEC Research and Development Report No. NYO-1480-82, New York Univ., Nov. 1967.

9. S. Chandrasekhar, *Hydrodynamic and Hydromagnetic Stability*, Internat. Series of Monographs on Physics, Clarendon Press, Oxford, 1961. MR **23** #1270.

10. A. J. Chorin, "Numerical study of thermal convection in a fluid layer heated from below," AEC Research and Development Report No. NYO-1480-61, New York Univ., Aug. 1966.

11. P. H. Rabinowitz, "Nonuniqueness of rectangular solutions of the Benard problem," *Arch. Rational Mech. Anal.* (To appear.)

12. E. L. Koschmieder, "On convection on a uniformly heated plane," *Beitr. Physik. Atm.*, v. 39, 1966, p. 1.

13. H. T. Rossby, "Experimental study of Benard convection with and without rotation," Ph.D. Thesis, Massachusetts Institute of Technology, Cambridge, Mass., 1966.

14. F. Busse, "On the stability of two dimensional convection in a layer heated from below," *J. Math. Phys.*, v. 46, 1967, p. 140.

On the Convergence of Discrete Approximations to the Navier-Stokes Equations*

By Alexandre Joel Chorin

Abstract. A class of useful difference approximations to the full nonlinear Navier-Stokes equations is analyzed; the convergence of these approximations to the solutions of the corresponding differential equations is established and the rate of convergence is estimated. ■

Introduction. The Navier-Stokes equations, describing the motion of a viscous incompressible fluid, can be written in the dimensionless form

$$\partial_t v_i + \partial_i p = -v_j \partial_j v_i + \nabla^2 v_i + E_i, \qquad (\nabla^2 \equiv \sum \partial_j^2) \tag{1}$$

$$\operatorname{div} \mathbf{v} = 0 \tag{2}$$

where the vector $\mathbf{v}$, with components v_i, $i = 1, 2, 3$, is the velocity, p is the pressure, $\mathbf{E}$ is the external force, ∂_t denotes differentiation with respect to the time t and ∂_i denotes differentiation with respect to the space variable x_i, $i = 1, 2, 3$. Vector quantities are denoted by bold-face characters and the summation convention applies to the index j.

When a solution of these equations is required in some bounded domain Ω with boundary $\partial\Omega$, use is generally made of an appropriate difference approximation. A new class of such approximations was introduced and utilized in [1] and [2]; it is the purpose of this paper to establish the convergence of the solutions of such approximations to the solutions of Eqs. (1) and (2) in Ω.

To our knowledge, the first convergence proof for a difference approximation to the complete system (1) and (2) was given by Krzywicki and Ladyzhenskaya (see e.g. [3]). Their proof gives both more and less than the numerical analyst requires. It gives more because it actually establishes the existence of a certain weak solution of the equations. It gives less because it provides no estimate of the error and because it applies to a scheme which is not readily applied in practical calculation. Proofs related to that of Krzywicki and Ladyzhenskaya have been given by Temam [4], [5], for schemes which are as yet untested in practice.

In the present paper we shall adopt a different point of view. We shall assume that the differential equations have a solution with a certain number of continuous derivatives. Armed with this knowledge, we shall study difference schemes which are not merely usable, but even efficient. The methods analyzed are based on the following observations: Equation (1) can be written in the form

$$\partial_t \mathbf{v} + \operatorname{grad} p = \mathfrak{F}\mathbf{v} + \mathbf{E} \tag{1'}$$

Received September 12, 1968, revised November 3, 1968.
* This work was partially supported by the U.S. Atomic Energy Commission, Contract No. AT(30-1)-1480.

Reprinted from "On the Convergence of Discrete Approximations to the Navier-Stokes Equations," *Mathematics of Computation*, (1969), Volume 23, Pages 341–353, by permission of the American Mathematical Society.

where the vector $\mathfrak{F}\mathbf{v}$, with components $-v_j\partial_j v_i + \nabla^2 v_i$, is a functional of $\mathbf{v}$; Eq. (2) can be differentiated to yield

$$\text{div}\,(\partial_t \mathbf{v}) = 0\,. \tag{2'}$$

(1′) can therefore be written in the form

$$\partial_t \mathbf{v} = \mathcal{P}(\mathfrak{F}\mathbf{v} + \mathbf{E}) \tag{3}$$

where $\mathcal{P}$ is an orthogonal projection operator which projects vectors in $L_2(\Omega)$ onto the subspace of vectors with zero divergence in Ω and satisfying an appropriate boundary condition on $\partial\Omega$ (see e.g. [6], [7]). Usually the appropriate boundary condition is that the normal component of $\mathbf{v}$ vanishes. On the basis of these remarks the following procedure is followed: The time t is discretized; at every time level $\mathfrak{F}\mathbf{v}$, then $\mathcal{P}(\mathfrak{F}\mathbf{v} + \mathbf{E})$ are evaluated; this yields an approximation to $\partial_t\mathbf{v}$ which is used to obtain $\mathbf{v}$ at the next time level.

As will become apparent in the course of this work, the author has not obtained results as general as he may have wished. A convergence proof in both the maximum and L_2 norms, with a suitable error estimate, has been obtained only for the special problems in which the boundary conditions are replaced by periodicity conditions. This proof is presented in the next two sections; first the discrete analogues of the operators grad, div and $\mathcal{P}$ are described and studied; these operators are then used to present and analyze a difference scheme for the periodic initial value problem. The mixed initial value-boundary value problem is briefly discussed in a final section.

Preliminaries; The Operators D, G and P. We assume that Eqs. (1) and (2) have a solution $\mathbf{v}$, p, periodic in all spatial directions; without loss of generality in the proofs the periods can be taken equal to 1. Let l be the number of space dimensions; Ω is then the cube $0 \leqq x_i \leqq 1$, $i = 1, \cdots, l$. We cover Ω by a rectangular grid and assume that the mesh-widths in all directions are equal to the same small number h. The set of all mesh-nodes is denoted by Ω_h; Ω_h^0 is the set of nodes in the interior of Ω and $\partial\Omega_h$ is the set of nodes on the boundary of Ω. $\Omega_h^0 + \partial\Omega_h = \Omega_h$. $N = h^{-1} + 1$ is the number of mesh-points in each space direction.

Let f be a scalar function and let $\mathbf{u}$ be a vector function with components u_i, defined at the points of Ω_h. Let $z = [q, r]$ (if $l = 2$) or $z = [q, r, s]$ (if $l = 3$) be a point in Ω_h with coordinates qh, rh (qh, rh, sh if $l = 3$). The values of f, u_i at z are denoted by f_z, $u_{i(z)}$ or $f_{q,r}$, $u_{i(q,r)}$ ($f_{q,r,s}$, $u_{i(q,r,s)}$, if $l = 3$). The periodicity conditions become $f_{q+N-1,r} = f_{q,r}$ etc.

The inner product is defined for scalar functions f, g, by

$$(f, g) = \sum_{z\in\Omega_h^0} f_z g_z h^l + \frac{1}{2} \sum_{z\in\partial\Omega_h} f_z g_z h^l$$

and for vectors $\mathbf{u}$, $\mathbf{v}$ by

$$(\mathbf{u}\cdot\mathbf{v}) = \sum_{i=1}^{l} (u_i, v_i)\,,$$

where only half the vertices are counted in the boundary sums. As usual, we set

$$\|f\| = ((f, f))^{1/2}\,, \qquad \|\mathbf{u}\| = ((\mathbf{u}, \mathbf{u}))^{1/2}\,.$$

The shift operators $S_{\pm i}$ are defined by

$$S_{\pm 1} f_{q,r} = f_{q\pm 1,r}$$

$$S_{\pm 2} f_{q,r} = f_{q,r\pm 1}$$

with similar definitions in the three-dimensional case. The difference operators D_{+i}, D_{-i}, D_{0i} are defined by

$$D_{+i} = (S_{+i} - I)/h ,$$

$$D_{-i} = (I - S_{-i})/h ,$$

$$D_{0i} = (D_{+i} + D_{-i})/2 = (S_{+i} - S_{-i})/2h ,$$

where I is the identity. D_{+i}, D_{-i} and D_{0i} are respectively the forward, backward and centered difference operators in the ith direction.

Let D and $\mathbf{G}$ denote respectively the discrete approximations to the operators div and grad. Both D and $\mathbf{G}$ employ centered differences, i.e. for a vector $\mathbf{u}$ on Ω_h we set $D\mathbf{u} \equiv D_{0j} u_j$ and for a scalar function ϕ we set $G_i\phi \equiv D_{0i}\phi$. With these definitions, the following identities can be readily verified:

(4) $$(D\mathbf{u}, e) = 0 ,$$

where $e \equiv 1$ at all points of Ω_h, and

(5) $$(D\mathbf{u}, \phi) + (\mathbf{u}, \mathbf{G}\phi) = 0$$

for all $\mathbf{u}$ and ϕ. These are the analogues of the identities

$$\int_\Omega \operatorname{div} u \, dx = 0$$

and

$$\int_\Omega \phi \operatorname{div} \mathbf{u} dx + \int_\Omega \mathbf{u} \operatorname{grad} \phi dx = \int_\Omega \operatorname{div} \phi \mathbf{u} dx = 0$$

which hold for smooth periodic functions $\mathbf{u}$ and ϕ on Ω. For $\mathbf{u}$, ϕ periodic and three times continuously differentiable, we have

$$\|\mathbf{G}\phi - \operatorname{grad} \phi\| = O(h^2) , \qquad \|D\mathbf{u} - \operatorname{div} \mathbf{u}\| = O(h^2) .$$

We shall now discuss some consequences of our systematic use of centered differences. Let ψ be a function on Ω_h, let z_-, z_+ be two points a distance $2h$ (modulo 1) apart, and let z_0 be the point on the line joining z_- and z_+ and at a distance h from each. One of the components of $\mathbf{G}\psi$ at z_0 is a linear combination of ψ_{z_+} and ψ_{z_-}; we describe this situation by saying the z_+ and z_- are G-connected. We say that points z, z' belong to the same G-chain if there exist points $z_1, z_2, \cdots, z_n$ such that any two successive points in the sequence

$$z, z_1, z_2, \cdots, z_n, z'$$

are G-connected. Clearly Ω_h is the union of some number L of disjoint G-chains. If N is even $L = 1$; if N is odd $L = 2^l$. (Had we allowed the numbers of mesh-points in the several space directions to differ from each other, we would have found that

$L = 2^i$, i between 0 and l.) The following facts can now be verified: (i) when $\mathbf{G}\phi$ is given, ϕ is determined only up to L arbitrary constants; and (ii) the sum on the left-hand side of the identity (4) can be separated into L partial sums, each vanishing separately. We refrain from assuming that N is even and $L = 1$ so that our discussion remain valid for the nonperiodic case where $L \neq 1$ for all N; see the last section of this paper.

We are now ready to prove the following discrete analogue of a well-known decomposition theorem:

THEOREM 1. *Let* $\mathbf{u}$ *be a vector on* Ω_h *satisfying the periodicity conditions; then there exist a unique periodic vector* $\mathbf{u}^D$ *and a periodic function* ϕ *such that*

$$D\mathbf{u}^D = 0 \tag{6}$$

$$\mathbf{u} = \mathbf{u}^D + \mathbf{G}\phi \tag{7}$$

at all points of Ω_h, *with*

$$(\mathbf{u}^D, \mathbf{G}\phi) = 0 \,. \tag{8}$$

Proof. If $\mathbf{u}^D$ satisfying Eq. (6) exists, then Eq. (8) is clearly satisfied because $(\mathbf{u}^D, \mathbf{G}\phi) = -(D\mathbf{u}^D, \phi) = 0$.

We already know that ϕ in Eq. (7) can be determined only up to L arbitrary constants. To lift this indeterminacy we can impose L additional conditions; for example, we can number the G-chains and require that

$$\sum_i \phi_z = 0 \,, \qquad i = 1, \cdots, L \,, \tag{9}$$

where $\sum_i$ denotes summation over the ith G-chain.

The theorem is proved by verification of the Fredholm alternative. Let q_0 be the number of points in $\Omega_h{}^0$ and q_∂ the number of points on $\partial\Omega_h$ ($q_0 + q_\partial = N^l$). There are $q_0 + q_\partial/2$ values of ϕ and $l(q_0 + q_\partial/2)$ components of $\mathbf{u}^D$ to determine. Equation (6) represents $(q_0 + q_\partial/2)$ equations related by the L identities (4), i.e. $q_0 + q_\partial/2 - L$ independent equations. Equation (7) represents $l(q_0 + q_\partial/2)$ relations; together with Eqs. (9) the number of equations equals the number of unknowns.

Squaring (7) and using (8) we obtain

$$\|\mathbf{u}\|^2 = \|\mathbf{u}^D\|^2 + \|\mathbf{G}\phi\|^2 \,. \tag{10}$$

Therefore, if $\mathbf{u} = 0$, then $\mathbf{u}^D = 0$, $\mathbf{G}\phi = 0$ and $\phi = 0$; this proves the theorem.

Let H be the space of periodic vectors defined on Ω_h, let H_D be the subspace of periodic vectors $\mathbf{v}$ such that $D\mathbf{v} = 0$, and let H_G be the subspace of vectors of the form $\mathbf{G}\phi$, ϕ periodic; Theorem 1 states that H_G and H_D are orthogonal to each other, and that H is their direct sum. Let P be the orthogonal projection projecting H on H_D; (7) can be written in the form

$$\mathbf{u} = P\mathbf{u} + \mathbf{G}\phi \,.$$

We obviously have for all $\mathbf{u}$

$$\|P\mathbf{u}\| \leqq \|\mathbf{u}\| \,. \tag{11}$$

P is the discrete analogue of $\mathcal{P}$ (see Eq. (3)). Given $\mathbf{u}$, it is a fairly simple matter to evaluate $P\mathbf{u}$; efficient methods for so doing were described in [2]. For the sake of completeness, we exhibit a method for finding $P\mathbf{u}$ which, albeit inefficient, has the merit of simplicity. Consider the iteration system

$$\mathbf{w}^{m+1} = \mathbf{u} - \mathbf{G}\phi^m \qquad m \geqq 1 \,,$$
$$\phi^{m+1} = \phi^m - \theta D\mathbf{w}^{m+1} \qquad m \geqq 1 \,, \qquad \theta > 0$$

where $\mathbf{w}^m$, ϕ^m, $m \geqq 1$, are periodic and θ is a parameter. It is readily verified that for $0 < \theta < h^2/l^2$, $\mathbf{w}^m$ converges to $P\mathbf{u}$ as m increases for all initial guesses ϕ^1.

To conclude this section, we prove a number of inequalities which will be needed in later sections. We start with a discrete analogue of the Poincaré inequality. Consider the case $l = 2$. Let ψ be a function defined on Ω_h, and let $z = [p, q]$, $z' = [p', q']$ be two points on the ith G-chain; $p' = p + 2m_1$, $q' = q + 2m_2$. We have,

$$\psi_{p',q'} - \psi_{p,q} = \sum_{k=0}^{m_1-1} (G_1\psi)_{p+1+2k,q}\cdot 2h + \sum_{k=0}^{m_2-1} (G_2\psi)_{p',q+1+2k}\cdot 2h \,.$$

Therefore

$$|\psi_{p',q'} - \psi_{p,q}|^2 \leqq 4\left[\sum_{k=1}^{N-1} |G_1\psi|_{k,q}h + \sum_{k=1}^{N-1} |G_2\psi|_{p',k}h\right]^2$$
$$\leqq 8\left[\sum_{k=1}^{N-1} |G_1\psi|^2_{k,q}h + \sum_{k=1}^{N-1} |G_2\psi|^2_{p',k}h\right],$$

where the relation $(N-1)h = 1$ is used. We multiply both sides by h^4 and sum over all $[p, q]$ and $[p', q']$ in the ith G-chain, giving points on $\partial\Omega_h$ the weight $\frac{1}{2}$, and obtain

$$\sum_i h^2 \sum_i |\psi|^2_{p,q}h^2 + \sum_i h^2 \sum_i |\psi|^2_{p',q'}h^2 - 2\sum_i \psi_{p,q}h^2 \sum_i \psi_{p',q'}h^2$$
$$\leqq 8\sum_i h^3\|\mathbf{G}\psi\|^2 \leqq 8\|\mathbf{G}\psi\|^2 \,,$$

where $\sum_i$ denotes summation over the ith G-chain.

Let N_i be the number of points in the ith G-chain. We have

$$N_i \geqq (N-1)^2/L > \tfrac{1}{2}N^2/L \,.$$

Therefore

$$\sum_i h^2 \geqq \frac{1}{2L}$$

and

$$\frac{1}{L}\sum_i |\psi|_{p,q}h^2 \leqq 2\Big(\sum_i \psi_{p,q}h^2\Big)^2 + 8\|\mathbf{G}\psi\|^2 \,.$$

Summing over all G-chains we obtain

$$\|\psi\|^2 \leqq 2L \sum_{G\text{-chains}} \Big(\sum_i \psi_{p,q}h^2\Big)^2 + C_1\|\mathbf{G}\psi\|^2 \,, \tag{12}$$

where $C_1 = 8L^2$. A similar inequality can be derived in the three-dimensional case, with $C_1 = 12L^2$. The inequality (12) can now be used to prove the following theorem:

THEOREM 2. *Let* $\mathbf{u}$ *be a periodic vector defined on* Ω_h. *Then the following inequality holds*

$$\|\mathbf{u} - P\mathbf{u}\| \leqq \sqrt{C_1}\|D\mathbf{u}\| , \tag{13}$$

where $\sqrt{C_1}$ *is a constant independent of* $\mathbf{u}$ *and* h.

Proof. By Theorem 1, $\mathbf{u} - P\mathbf{u} = \mathbf{G}\phi$ is in H_G. Let $\mathbf{G}\psi$ be an arbitrary unit vector in H_G, $(\|\mathbf{G}\psi\| = 1)$. ψ is determined only up to L arbitrary constants which can be chosen so that

$$\sum_i \psi_z = 0 , \qquad i = 1, \cdots, L .$$

We have

$$(\mathbf{G}\phi, \mathbf{G}\psi) = (\mathbf{u} - P\mathbf{u}, \mathbf{G}\psi) = (\mathbf{u}, \mathbf{G}\psi) = -(D\mathbf{u}, \psi) .$$

Hence, using (12), we obtain

$$|(\mathbf{G}\phi, \mathbf{G}\psi)| = |(D\mathbf{u}, \psi)| \leqq \|D\mathbf{u}\| \, \|\psi\| \leqq \sqrt{C_1}\|D\mathbf{u}\| .$$

Since $\mathbf{G}\psi$ is an arbitrary unit vector in H_G, (13) follows.

Solution of the Periodic Initial-Value Problem. In this section a scheme for finding periodic solutions of Eqs. (1) and (2) will be analyzed. The particular scheme discussed has been singled out because it resembles schemes the author has used in actual computation (see [2]); it will be evident that the analysis applies to wide classes of schemes. We shall again simplify notations by writing the equations for the two-dimensional case; the scheme as well as the proofs generalize to the three-dimensional case without further ado.

Let $\mathbf{u}$, with components u_i, be the computed velocity, let π be the computed pressure, and let k be the time step. We write

$$\mathbf{u}^n \equiv \mathbf{u}(nk) , \qquad \pi^n \equiv \pi(nk) , \qquad \text{etc.}$$

At the time $t = 0$ a periodic velocity field $\mathbf{u}^0$ is assumed given. (More will be said later about the proper choice of $\mathbf{u}^0$.) Given $\mathbf{u}^n$, $\mathbf{u}^{n+1}$ is evaluated in three fractional steps:

$$u_i^{n+1/3} = u_i^n - ku_1^n D_{01} u_i^{n+1/3} + kD_{+1}D_{-1}u_i^{n+1/3} \tag{14a}$$

$$u_i^{n+2/3} = u_i^{n+1/3} - ku_2^n D_{02} u_i^{n+2/3} + kD_{+2}D_{-2}u_i^{n+2/3} \tag{14b}$$

$$\mathbf{u}^{n+1} = P(\mathbf{u}^{n+2/3} + k\mathbf{E}^{n+1}) \tag{14c}$$

with $\mathbf{u}^{n+1/3}$, $\mathbf{u}^{n+2/3}$ periodic.

Equations (14a) and (14b) can be rewritten in the form

$$(I - kQ_1(\mathbf{u}^n))\mathbf{u}^{n+1/3} = \mathbf{u}^n , \tag{15a}$$

$$(I - kQ_2(\mathbf{u}^n))\mathbf{u}^{n+2/3} = \mathbf{u}^{n+1/3} , \tag{15b}$$

where $Q_1(\mathbf{u}^n)$, $Q_2(\mathbf{u}^n)$, are linear operators dependent on the parameters $u_{i(z)}^n$. Equation (14c) can be rewritten in the form

$$\mathbf{u}^{n+1} + k\mathbf{G}\pi^{n+1} = \mathbf{u}^{n+2/3} + k\mathbf{E}^{n+1} \qquad (D\mathbf{u}^{n+1} = 0) \tag{15c}$$

which defines π^{n+1}, the computed pressure at the time $(n+1)k$. ($\mathbf{u}^{n+2/3}$ corresponds to $\mathbf{u}^{\text{aux}}$ in the notations of [1] and [2].) It can be seen that the vector $(\mathbf{u}^{n+2/3} - \mathbf{u}^n)/k$ approximates $\mathfrak{F}\mathbf{u}$ and that Eq. (14c) which is equivalent to

$$(\mathbf{u}^{n+1} - \mathbf{u}^n)/k = P\{(\mathbf{u}^{n+2/3} - \mathbf{u}^n)/k + \mathbf{E}\}$$

is the discrete analogue of Eq. (3).

The task now at hand is to prove that $\mathbf{u}^{n+1/3}$, $\mathbf{u}^{n+2/3}$, $\mathbf{u}^{n+1}$ exist, i.e. that the operators $(I - kQ_i(\mathbf{u}^n))$ are invertible when $\mathbf{u}^0$ is chosen appropriately, and that the vectors $\mathbf{u}^n$ converge to the solution $\mathbf{v}(nk)$ of Eqs. (1) and (2). We start by showing that Eqs. (14) are consistent with Eqs. (1) and (2); this is the content of the following lemma:

LEMMA 1. *Let $k = O(h^2)$, and assume that Eqs.* (1) *and* (2) *have a periodic solution* $\mathbf{v}$, p, *which has continuous derivatives up to order five in the interval* $0 \leqq t \leqq T$. *Then there exist two times continuously differentiable vectors* $\mathbf{w}^n$, $\mathbf{w}^{n+1/3}$, $\mathbf{w}^{n+2/3}$ $(0 \leqq nk < T)$ *such that*

$$(16a)\qquad (I - kQ_1(\mathbf{w}^n))\mathbf{w}^{n+1/3} = \mathbf{w}^n + O(k^2)\,,$$

$$(16b)\qquad (I - kQ_2(\mathbf{w}^n))\mathbf{w}^{n+2/3} = \mathbf{w}^{n+1/3} + O(k^2)\,,$$

$$(16c)\qquad \mathbf{w}^{n+1} = P(\mathbf{w}^{n+2/3} + k\mathbf{E}^{n+1}) + O(k^2)\,,$$

with

$$(17)\qquad \|\mathbf{v}^n - \mathbf{w}^n\| = O(k)\,.$$

Proof. We simply construct the required functions. We have

$$D\mathbf{v}^n = \operatorname{div}\mathbf{v}^n + \sum_{\beta=1}^{l} \frac{h^2}{3!}\,\partial_\beta{}^3 v_\beta{}^n + O(h^4)\,.$$

Therefore, putting

$$(18)\qquad w_\beta{}^n = v_\beta{}^n - \frac{h^2}{3!}\,\partial_\beta{}^2 v_\beta{}^n \qquad \beta = 1, \cdots, l \text{ (no summation over } \beta)$$

we obtain $D\mathbf{w}^n = O(h^4) = O(k^2)$. Equation (17) is clearly satisfied, and by Theorem 2

$$(19)\qquad \|\mathbf{w}^n - P\mathbf{w}^n\| = O(k^2)\,.$$

We now put

$$w_i^{n+1/3} = w_i^{n+1} - k\partial_2{}^2 v_i^{n+1} + kv_2{}^n\partial_2 v_i^{n+1} + k\partial_i p^{n+1} - kE_i^{n+1}$$

$$w_i^{n+2/3} = w_i^{n+1} + k\partial_i p^{n+1} - kE_i^{n+1}\,.$$

Equations (16a) and (16b) are clearly satisfied, and since

$$\mathbf{G}p = \operatorname{grad} p + O(h^2) = \operatorname{grad} p + O(k)$$

we have

$$P(k \operatorname{grad} p) = O(k^2)\,.$$

On the other hand, it can always be assumed that $\operatorname{div}\mathbf{E} = 0$. (Since adding a gradient to $\mathbf{E}$ merely changes the definition of p; see e.g. [7]), and therefore by Theorem 2

$$\|k\mathbf{E} - kP\mathbf{E}\| \leqq \sqrt{C_1 k}\|D\mathbf{E}\| = O(k^2)\,.$$

These equations, together with Eq. (19) show that (16c) is satisfied, and the lemma is proved.

We shall use $\mathbf{w}^n$ as a comparison vector, i.e. we shall prove that $\|\mathbf{u}^n - \mathbf{w}^n\|$ is small, and use (17) at the end of the argument to show that $\|\mathbf{u}^n - \mathbf{v}^n\|$ is small. The lemma assumes that $\mathbf{v}$ has continuous derivatives up to order five. Had we assumed the existence of only four continuous derivatives, the error term in (16c) would have been of order kh. This is sufficient for convergence; however, the proof becomes somewhat more complicated and we shall be content with the assumption of the lemma.

We now introduce a second norm, the discrete maximum norm, defined for scalar functions ϕ by

$$\|\phi\|_{\max} = \max_{z \in \Omega_h} |\phi_z|$$

and for vectors $\mathbf{u}$ by

$$\|\mathbf{u}\|_{\max} = \max_i \|u_i\|_{\max}\,.$$

We have

LEMMA 2. $\|\phi\|_{\max} \leqq h^{-l/2}\|\phi\|$, $\|\mathbf{u}\|_{\max} \leqq h^{-l/2}\|\mathbf{u}\|$.

The proof is obvious from the definitions. This lemma is crucial to the sequel since, as we shall see, it implies that if $\mathbf{u}$ converges to $\mathbf{v}$ with sufficient accuracy in the L_2 norm, then $\mathbf{u}$ also converges to $\mathbf{v}$ in the maximum norm.

LEMMA 3. *Let $\|\mathbf{u}^n\|_{\max} \leqq K$, and let k be small enough for the inequality $kK^2/4 < 1$ to hold. Equations* (14a) *and* (14b) *can then be uniquely solved for $\mathbf{u}^{n+1/3}$, $\mathbf{u}^{n+2/3}$.*

Proof. Multiplying (14a) by $u_i^{n+1/3}$ we obtain

$$\begin{aligned}\|u_i^{n+1/3}\|^2 = &- k(u_i^{n+1/3}, u_1^n D_{01} u_i^{n+1/3}) \\ &+ k(u_i^{n+1/3}, D_{+1}D_{-1}u_i^{n+1/3}) \\ &+ (u_i^n, u_i^{n+1/3})\,,\end{aligned}$$

however, we have

$$(u_i^{n+1/3}, D_{+1}D_{-1}u^{n+1/3}) = -\|D_{+1}u_i^{n+1/3}\|^2$$

$$|(u_i^{n+1/3}, u_1^n D_{01}u_i^{n+1/3})| \leqq \|u_1^n\|_{\max}\|u_i^{n+1/3}\|\,\|D_{01}u_i^{n+1/3}\|\,.$$

On the other hand, $D_{01} = \frac{1}{2}(D_{+1} + D_{-1})$, therefore

$$\|D_{01}u_i^{n+1/3}\| \leqq \tfrac{1}{2}(\|D_{+1}u_i^{n+1/3}\| + \|D_{-1}u_i^{n+1/3}\|) = \|D_{+1}u_i^{n+1/3}\|$$

and

$$\begin{aligned}|k(u_i^{n+1/3}, u_1^n D_{01}u_i^{n+1/3})| &\leqq Kk\|u_i^{n+1/3}\|\,\|D_{+1}u_i^{n+1/3}\| \\ &\leqq k\|D_{+1}u_i^{n+1/3}\|^2 + (K^2k/4)\|u_i^{n+1/3}\|^2\,,\end{aligned}$$

and hence

$$\|u_i^{n+1/3}\|\left(1 - \frac{kK^2}{4}\right) \leqq \|u_i^n\|\,. \tag{20}$$

The existence and uniqueness of $u_i^{n+1/3}$ follow by the Fredholm alternative. The existence and uniqueness of $u_i^{n+2/3}$ are established in the same way.

LEMMA 4. *Let $\|\mathbf{u}^n\|_{\max} \leqq K$, and let k be small enough for the inequality $kK^2/2 < 1$ to hold; then, if $k = O(h^2)$, we have*

$$\|\mathbf{u}^{n+1} - \mathbf{w}^{n+1}\| \leqq (1 + kC_2(K))\|\mathbf{u}^n - \mathbf{w}^n\| + C_3kh^2 , \tag{21}$$

where C_3 is a constant, and $C_2(K)$ is a constant whose magnitude depends on K.

Proof. Subtracting (14a) from (16a) we obtain

$$u_i^{n+1/3} - w_i^{n+1/3} = - ku_1^n D_{01}(u_i^{n+1/3} - w_i^{n+1/3}) - k(u_1^n - w_1^n)D_{01}w_i^{n+1/3} + kD_{+1}D_{-1}(u_i^{n+1/3} - w_i^{n+1/3}) + O(k^2) .$$

Multiplication by $u_i^{n+1/3} - w_i^{n+1/3}$ and manipulations similar to those in the proof of Lemma 3 yield

$$\|u_i^{n+1/3} - w_i^{n+1/3}\|\left(1 - \frac{kK^2}{4}\right) \leqq \|u_i^n - w_i^n\|(1 + kM_1) + O(k^2) ,$$

where

$$M_1 = \max_i \max_{0\leqq t\leqq T} \max_\Omega |\partial_1 w_i^{n+1/3}| .$$

Similarly, we obtain

$$\|u_i^{n+2/3} - w_i^{n+2/3}\|\left(1 - \frac{kK^2}{4}\right) \leqq \|u_i^{n+1/3} - w_i^{n+1/3}\| + kM_2\|u_i^n - w_i^n\| + O(k^2) ,$$

where

$$M_2 = \max_i \max_{0\leqq t\leqq T} \max_\Omega |\partial_2 w_i^{n+2/3}|$$

and hence

$$\|\mathbf{u}^{n+2/3} - \mathbf{w}^{n+2/3}\| \leqq (1 + kC_2(K))\|\mathbf{u}^n - \mathbf{w}^n\| + O(k^2) ,$$

where $C_2(K)$ depends on K. Finally

$$\mathbf{u}^{n+1} - \mathbf{w}^{n+1} = P(\mathbf{u}^{n+2/3} - \mathbf{w}^{n+2/3}) + O(k^2)$$

and, therefore, using (11)

$$\begin{aligned}\|\mathbf{u}^{n+1} - \mathbf{w}^{n+1}\| &\leqq \|\mathbf{u}^{n+2/3} - \mathbf{w}^{n+2/3}\| + O(k^2) \\ &\leqq (1 + kC_2(K))\|\mathbf{u}^n - \mathbf{w}^n\| + C_3kh^2\end{aligned}$$

and the lemma is proved.

Let $\mathbf{u}^0$ be the initial value of $\mathbf{u}$, for use in Eqs. (14). We assume that

$$\|\mathbf{u}^0 - \mathbf{w}^0\| = C_4h^2 . \tag{22}$$

This can be achieved for example by putting $\mathbf{u}^0 = \mathbf{v}^0$. Let W be defined by

$$W = \max_i \max_{0\leqq t\leqq T} \max_\Omega |w_i| .$$

Let C be the largest of C_3, C_4 and $C_2(2W)$. Assume k is so small that $kC << 1$ and that h is smaller than h_0, where

$$\max(C, 1)h_0^{(4-l)/2} = \tfrac{1}{2}\epsilon W, \qquad \epsilon < 1.$$

Equation (22) and Lemma 2 then show that

$$\|\mathbf{u}^0\|_{\max} \leqq W + \tfrac{1}{2}\epsilon W < 2W.$$

By Lemma 3 $\mathbf{u}^1$ exists and by Lemma 4 we have $\|\mathbf{u}^1 - \mathbf{w}^1\| \leqq (1 + Ck)Ch^2 + Ckh^2$. Therefore

$$\|\mathbf{u}^1 - \mathbf{w}^1\|_{\max} \leqq \frac{\epsilon}{2}(1 + Ck)W + \frac{\epsilon}{2}kW < \epsilon W < W, \qquad \|\mathbf{u}^1\|_{\max} < 2W,$$

and we can evaluate $\mathbf{u}^2$. In general we have

$$\begin{aligned} \|\mathbf{u}^{n+1} - \mathbf{w}^{n+1}\| &\leqq (1 + Ck)^{n+1}Ch^2 + [1 + (1 + Ck) + \cdots + (1 + Ck)^n]Ckh^2 \\ &< 2e^{C(n+1)k}\max(C, 1)h^2 \end{aligned} \tag{23}$$

and

$$\|\mathbf{u}^{n+1} - \mathbf{w}^{n+1}\|_{\max} \leqq \epsilon W e^{Ct}, \qquad (t = (n + 1)k). \tag{24}$$

Let T_0 be defined by $\exp CT_0 = 1/\epsilon$ and let $T_1 = \min(T, T_0)$. Inequality (24) shows that for $0 \leqq t \leqq T_1$, $\|\mathbf{u}\|_{\max} \leqq 2W$ and hence for $0 \leqq nk < T_1$, $\mathbf{u}^{n+1}$ exists and

$$\|\mathbf{u}^{n+1} - \mathbf{w}^{n+1}\| \leqq 2\max(C, 1)e^{Ct}h^2 \tag{25a}$$

as well as

$$\|\mathbf{u}^{n+1} - \mathbf{w}^{n+1}\|_{\max} \leqq 2\max(C, 1)e^{Ct}h^{(4-l)/2}. \tag{25b}$$

If $T_1 < T$, i.e. if the inequalities (25) hold for a time interval shorter than the time interval for which the solution of the differential equations has five bounded derivatives and for which a numerical solution is required, the above process can be restarted at $t = T_1$, to yield convergence for the whole finite interval $0 \leqq t \leqq T$.

Bearing in mind the definition of $\mathbf{w}$ and Eq. (17), we obtain the following theorem:

THEOREM 3. *Let Eqs.* (1) *and* (2) *have a periodic solution with continuous derivatives up to order five for* $0 \leqq t \leqq T$. *Let* $k = O(h^2)$; *if* $\|\mathbf{u}^0 - \mathbf{w}^0\|$, *k and h are sufficiently small, Eqs.* (14) *have a unique solution which converges to the solution of* (1) *and* (2) *in both the* L_2 *and maximum norms. The error in the* L_2 *norm is of order* h^2; *the error in the maximum norm is bounded by* $O(h)$ *in the two-dimensional case and by* $O(\sqrt{h})$ *in the three-dimensional case.*

Theorem 3 and its proof can be summarized as follows: Let $\mathbf{u}_z{}^n$, $\mathbf{w}_z{}^n$ be vector functions defined for z in Ω_h and for n such that $0 \leqq nk \leqq T_1$; introduce the "space-time" maximum and L_2 norms

$$\|\mathbf{u}\|_{\max, T_1} = \max_{0 \leqq nk < T_1} \|\mathbf{u}^n\|_{\max}$$

$$\|\mathbf{u}\|_{T_1} = \max_{0 \leqq nk < T_1} \|\mathbf{u}^n\|.$$

The equations

$$(I - kQ_1(\boldsymbol{\omega}^n))\mathbf{u}^{n+1/3} = \mathbf{u}^n ,$$
$$(I - kQ_2(\boldsymbol{\omega}^n))\mathbf{u}^{n+2/3} = \mathbf{u}^{n+1/3} ,$$
$$\mathbf{u}^{n+1} = P(\mathbf{u}^{n+2/3} + k\mathbf{E}^{n+1}) ,$$
$$\mathbf{u}^0 \text{ given} ,$$

define a mapping $\boldsymbol{\omega} \to \mathbf{u}$. This mapping maps the maximum norm sphere

$$\|\boldsymbol{\omega}\|_{\max,T_1} \leqq 2\|\mathbf{w}\|_{\max,T_1}$$

into the L_2 norm sphere

$$\|\mathbf{u} - \mathbf{w}\|_{T_1} \leqq \|\mathbf{w}\|_{\max,T_1} h^{l/2} .$$

For $\|\mathbf{u}^0 - \mathbf{w}^0\|$, k and h sufficiently small, this mapping has a unique fixed point which is the solution of (14) and lies close to $\mathbf{v}$, the solution of (1) and (2).

In our analysis we have neglected the effect of round-off error and of the errors arising from the possibly incomplete iterative evaluation of $\mathbf{u}^{n+1} = P(\mathbf{u}^{n+2/3} + k\mathbf{E}^{n+1})$. It is obvious, however, that the analysis remains valid if the round-off errors are of order k^2 and provided $\mathbf{u}^{n+1}$ is approximated by a vector $(\mathbf{u}^{n+1})^*$ such that $D(\mathbf{u}^{n+1})^* = O(k^2)$. Furthermore, in the dimensionless variables used in this paper the effect of the Reynolds number R on the error is not in evidence. Clearly C depends on R and increases as R increases; i.e. as R increases k and h have to be reduced for accuracy to be preserved. Finally, it is clear that the results of this section apply to certain other quasi-linear equations besides the Navier-Stokes equations, provided the boundary conditions are homogeneous. In this sense, our results generalize the work of M. Lees (see e.g. [8]), who considered equations with nonlinear terms of a simpler nature.

The Mixed Initial Value-Boundary Value Problem. The main interest of methods such as those considered in this paper lies in their applicability to mixed initial value-boundary value problems. Schemes similar to (14) have been successfully applied by the author to a variety of such problems (see. e.g. [2]). The convergence proof however, becomes considerably more difficult in the presence of boundaries.

Consider in particular the problem of solving Eqs. (1) and (2) in a domain Ω, with $\mathbf{v}^0$ given and with the boundary condition

(26) $$\mathbf{v} = 0 \quad \text{on } \partial\Omega .$$

Operators D and $\mathbf{G}$ can be constructed so that the identities (4) and (5) are satisfied and Theorems 1 and 2 hold. D and $\mathbf{G}$ thus constructed employ centered differences except on $\partial\Omega$. On $\partial\Omega$ one-sided first-order differences are used whenever the use of centered differences would require functional values at points outside Ω. The projection P associated with $\mathbf{G}$ and D is orthogonal in the space of functions satisfying (26). The proofs of all these statements take into account the fact that the number of G-chains is 2^l independently of the number of points in the mesh.

Difficulties arise, however, when one approaches the convergence proof proper. It is clear from the proof of Lemma 1 of the last section that, were one to use schemes such as (14), one would have to impose on $\mathbf{u}^{n+1/3}$, $\mathbf{u}^{n+2/3}$ inhomogeneous boundary conditions of the form

$$\mathbf{u}^{n+1/3} = k\mathbf{G}\pi^n \quad \text{on } \partial\Omega_h \tag{27a}$$

$$\mathbf{u}^{n+2/3} = k\mathbf{G}\pi^n \quad \text{on } \partial\Omega_h \tag{27b}$$

where π^n is the pressure computed at time $t = nk$. Such a procedure has indeed been followed in practice. Unfortunately, in the presence of inhomogeneous boundary conditions the author has not been able to establish the analogues of Lemmas 3 and 4. Moreover, the construction of $\mathbf{w}$ in Lemma 1 does not carry over to the present problem, since $\mathbf{w}$, as given in the last section, does not satisfy the imposed boundary conditions. Both difficulties stem from the fact that in the presence of boundaries the operators P and ∇^2 (linear part of $\mathfrak{F}$) do not commute. This is reminiscent of other situations in numerical analysis where the noncommutativity of certain operators hinders the analysis of fractional-step methods without detracting from their practical usefulness.

It is nevertheless possible to develop schemes for which convergence in the L_2 norm can be established. As an example, consider the following scheme with two fractional steps:

$$u_i^{n+1/2} = u_i^n - k\sum_\beta \{\tfrac{1}{2}(S_{+\beta} + S_{-\beta})(u_\beta^n D_{0\beta} u_i^{n+1/2}) + D_{+\beta}D_{-\beta}u_i^{n+1/2}\} \quad \text{in } \Omega_h{}^0 \tag{28a}$$

$$\mathbf{u}^{n+1/2} = 0 \quad \text{on } \partial\Omega_h \tag{28b}$$

$$\mathbf{u}^{n+1} = P(\mathbf{u}^{n+1/2} + k\mathbf{E}^{n+1})\,. \tag{28c}$$

It is clear the homogeneous boundary condition (28b) contains an error of order k. However, since the number of mesh-points on the boundary is $O(h)$ times the number of mesh-points in the whole domain, some accuracy in the L_2 norm will be preserved. We shall indicate how one can establish that in the L_2 norm the solution of (28) converges to the solution of the Navier-Stokes equations which satisfies the correct boundary conditions. $\mathbf{u}^n$, as given by (28), therefore assumes the imposed boundary conditions in a weak sense. It is clear that the estimates we shall derive will not do justice to the accuracy of the method.

One can verify the following identity

$$\left(f, \sum_\beta \frac{1}{2}(S_{+\beta} + S_{-\beta})(u_\beta^n D_{0\beta} f)\right) = 0$$

which holds for all f provided $D\mathbf{u} = 0$ in Ω_h and $\mathbf{u} = 0$ on the boundary. This of course is a discrete analogue of the identity $\int_\Omega f u_j \partial_j f dx = 0$, which holds whenever $\operatorname{div} \mathbf{u} = 0$ in Ω and $\mathbf{u} = 0$ on the boundary. Using this identity we can establish the following inequalities:

$$\|\mathbf{u}^{n+1/2}\| \leqq \|\mathbf{u}^n\|$$

and

$$\|\mathbf{u}^{n+1}\| \leqq \|\mathbf{u}^n\| + k\|P\mathbf{E}^{n+1}\| \leqq \|\mathbf{u}^0\| + k\sum_{i=0}^{n+1} \|P\mathbf{E}^i\|\,.$$

If we assume that Eqs. (1) and (2) have a solution $\mathbf{v}$ with continuous derivatives up to order four, this inequality can be used to show that if $k = O(h^2)$

$$\|\mathbf{u}^n - \mathbf{v}^n\| \leqq \text{constant} \sqrt{h}\,, \qquad 0 \leqq nk \leqq T\,. \tag{29}$$

For two-dimensional problems one can replace (28a) by an explicit scheme (which does not require intermediate boundary data such as (28b)). For small enough Reynolds number and provided $k < h^2/4$ one can then derive an estimate similar to (29). Furthermore, the scheme (28) can be modified so that a convergence proof of the Krzywicki-Ladyzhenskaya type becomes possible.

Since neither the scheme (28) nor its modifications are of any particular practical significance, details and proofs are omitted. (It should be pointed out however, that the system of linear equations (28a) can be solved by successive relaxation, provided the relaxation factor ω is sufficiently small. For proof, see [9].)

In ending, the author would like to make some comments on the preceding proofs. First of all, he would like to state his belief that the value of a scheme such as (14) lies in its practical usefulness, not in the possible existence of a convergence proof. The value of the convergence proofs lies in the fact that they contribute to the understanding of the numerical processes performed on the computer.

The proof of this paper requires the existence of four or five continuous derivatives of $\mathbf{v}$ and p. Furthermore, the error increases as the bounds on the required derivatives increase. This situation is inherent in the very nature of difference schemes; as a result, it is highly improbable that a flow containing a strong cascade process, i.e. a process in which energy is transferred from large to small eddies, can be adequately described by a difference method, for indeed, such flows are characterized by rapid increase in the higher derivatives. This of course excludes turbulence from the range of application of difference methods.

Finally, it has been claimed by several authors that the nonlinear terms in the Navier-Stokes equations must always be cast in "conservation law" form, i.e. in a form which implies the existence of identities for the momentum similar in appearance to those which hold for the solutions of the differential equations. The author knows of no good reason for following this procedure in problems with a smooth solution and has not endeavored to do so.

Courant Institute of Mathematical Sciences
New York University
251 Mercer Street
New York, New York 10012

1. A. J. Chorin, "The numerical solution of the Navier-Stokes equations for an incompressible fluid," *Bull. Amer. Math. Soc.*, v. 73, 1967, pp. 928–931. MR **35** #7643.

2. A. J. Chorin, "Numerical solution of the Navier-Stokes equations," *Math. Comp.*, v. 22, 1968.

3. A. Krzywicki & O. A. Ladyzhenskaya, "A grid method for the Navier-Stokes equations," *Soviet Phys. Dokl.*, v. 11, 1966, p. 212.

4. R. Temam, "Une méthode d'approximation de la solution des équations de Navier-Stokes," *Bull. Soc. Math. France.* (To appear.)

5. R. Temam, "Sur l'approximation de la solution des équations de Navier-Stokes par la méthode des pas fractionnaires," *Arch. Rational Mech. Anal.* (To appear.)

6. H. Fujita & T. Kato, "On the Navier-Stokes initial value problem," *Arch. Rational Mech. Anal.*, v. 16, 1964, pp. 269–315. MR **29** #3774.

7. O. A. Ladyzhenskaya, *Mathematical Problems in the Dynamics of a Viscous Incompressible Flow*, Fizmatgiz, Moscow, 1961; English transl., Gordon & Breach, New York, 1963. MR **27** #5034a, b.

8. M. Lees, "Energy inequalities for the solution of differential equations," *Trans. Amer. Math. Soc.*, v. 94, 1960, pp. 58–73. MR **22** #4875.

9. A. J. Chorin & O. Widlund, "On the convergence of relaxation methods." (To appear.)

Numerical Solution of Boltzmann's Equation*

ALEXANDRE JOEL CHORIN

1. Introduction

Boltzmann's equation describes the evolution of the one-particle distribution function $f = f(\mathbf{x}, \mathbf{u}, t)$, where the vector $\mathbf{x}$, with components x_1, x_2, x_3, is the position vector, $\mathbf{u}$, with components u_1, u_2, u_3, is the velocity vector, and t is the time. In the case of a gas of elastic sphere and in the absence of external forces, this equation takes the form

$$\frac{\partial f}{\partial t} + (\mathbf{u} \cdot \nabla) f = \tfrac{1}{2}\sigma^2 \int |\mathbf{V} \cdot \mathbf{e}| \, (f^+ f^{+\prime} - f f') \, d\omega \, d\mathbf{u}', \tag{1}$$

where ∇ is the gradient operator, σ is the radius of a particle, $\mathbf{e}$ is the unit vector pointing in the direction of the element of solid angle $d\omega$, and $\mathbf{V} = \mathbf{u}' - \mathbf{u}$ is the relative velocity upon impact of the particles with velocities $\mathbf{u}'$ and $\mathbf{u}$. The velocities $\mathbf{u}^+, \mathbf{u}^{+\prime}$ before collision of those spheres, which after collision acquire the velocities $\mathbf{u}, \mathbf{u}'$, are

$$\mathbf{u}^+ = \mathbf{u} + (\mathbf{V} \cdot \mathbf{e})\mathbf{e},$$

$$\mathbf{u}^{+\prime} = \mathbf{u}' - (\mathbf{V} \cdot \mathbf{e})\mathbf{e}.$$

The functions $f, f', f^+, f^{+\prime}$ in the integrand of the collision integral on the right-hand side of (1) are defined by

$$f = f(\mathbf{x}, \mathbf{u}, t),$$

$$f' = f(\mathbf{x}, \mathbf{u}', t),$$

$$f^+ = f(\mathbf{x}, \mathbf{u}^+, t),$$

$$f^{+\prime} = f(\mathbf{x}, \mathbf{u}^{+\prime}, t),$$

* The work presented in this paper was performed at the Courant Institute of Mathematical Sciences and was partially supported by the Atomic Energy Commission under Contract No. AT(30-1)-1480.

Reprinted with permission from *Communications on Pure and Applied Mathematics*, Vol. 25, 1972.

and the integration is carried out over all velocities $\mathbf{u}'$ and over the unit sphere. For an elementary discussion of equation (1), see [15]; for a thorough discussion see e.g., [2] and [6]. It is the purpose of this paper to present a numerical algorithm for solving this equation, and to apply it to the problem of shock structure in a one-dimensional flow. It will be seen that the method of solution generalizes to problems involving other molecular models and more space dimensions.

Unlike the work presented in the present paper, most previous numerical treatments of Boltzmann's equation relied on a Monte-Carlo technique; some of these treatments are ingenuous and interesting, but none can be considered accurate (see e.g., [1], [9], and [12]). Reference [9] is particularly clear and helpful.

For any function $\phi(\mathbf{x}, \mathbf{u})$, let $\bar{\phi}(x)$ denote the integral

$$\bar{\phi}(x) = \int \phi(\mathbf{x}, \mathbf{u}) f(\mathbf{x}, \mathbf{u}) \, d\mathbf{u} .$$

Some of the quantities of interest in the solution of the Boltzmann equation are the following moments of f: the density $\rho(\mathbf{x}) = \bar{1}$; the mean velocity $\mathbf{v} = \bar{\mathbf{u}}/\rho$; the pressure $p = \frac{1}{3}\overline{\mathbf{w}^2}$, where $\mathbf{w} = \mathbf{u} - \mathbf{v}$; the temperature $T = p/\rho R$, where R is the universal gas constant; the momentum transfer tensor $p_{ij} = \overline{w_i w_j}$, the w_i being the components of $\mathbf{w}$, and the heat flux vector $\mathbf{s} = \frac{1}{2}\overline{(\mathbf{w}^2)\mathbf{w}}$. Other quantities of interest are the Boltzmann H function

$$H = \int f \log f \, d\mathbf{u} ,$$

and, in shock problems, various geometrical parameters which characterize the structure of the shock.

2. Method of Solution

There are several major difficulties in the solution of equation (1). The function f depends on a relatively large number of independent variables, and the domain of f as a function of $\mathbf{u}$ is infinite, so that if (1) is replaced by a system of algebraic equations, their number will be large. The presence of the nonlinear collision integral ensures that the algebraic equations will be not only numerous, but also very cumbersome. There is a critical need for efficient computation. The integration over the angular variables introduces a further major hurdle. Suppose f is represented by a discrete set of values assumed on a discrete set Z of points in phase space. The integration over $\mathbf{u}'$ becomes a sum over the values assumed by f on Z. The angular integration will become a sum over a discrete set θ of angles. For any reasonable choice of Z and θ, the arguments of

f^+, $f^{+'}$ will include points not in Z. Thus interpolation, both accurate and stable, between values of f on Z will be required; for a discussion of the problem, see e.g., [17].

These difficulties can be resolved as follows: Once one resigns oneself to the need for interpolation, there is no reason to identify Z with the equidistant nodes of a regular mesh. In particular, one can make Z consist of points $(\mathbf{x}, \mathbf{u})$, where the components of $\mathbf{u}$ are roots of polynomials of degree n, $P_n(u)$, with $P_0, P_1, P_2, \cdots, P_n, \cdots$ forming an orthogonal sequence with respect to a weight $W(u)$. When $W(u)$ is properly chosen, f/W has a rapidly convergent series expansion in the polynomials $P_i(u)$, $i = 0, 1, \cdots$, and this expansion can be approximately constructed from the values of f on Z. It can then be used to provide the values of f outside Z. In the case of the Boltzmann equation, it is natural to use the weight function $W = \pi^{-1/2} e^{-u^2}$, and thus to identify the polynomials $P_n(u)$ with the Hermite polynomials $H_n(u)$,

$$H_n(u) = (-1)^n c_n e^{u^2} \frac{d^n}{du^n} e^{-u^2}, \qquad c_n = (2^n n!)^{-1/2};$$

the set $H_n(u)e^{-u^2/2}$ is complete on $L_2(-\infty, +\infty)$, and the $H_n(u)$ satisfy the orthogonality conditions

$$\pi^{-1/2} \int H_n(u) H_m(u) e^{-u^2} du = \delta_{n,m},$$

$\delta_{n,m}$ being the Kronecker delta (see e.g., [11]).

We are thus led to the following step-by-step procedure for solving equation (1): divide the time into segments of length Δt; assume that at time $t = n\Delta t$, n integer, f is given by a series

$$f(\mathbf{x}, \mathbf{u}, n\,\Delta t) = \pi^{-3/2}(s_1 s_2 s_3)^{-1/2} \sum_i \sum_j \sum_k a_{ijk}(\mathbf{x}, n\,\Delta t) \times H_i(\alpha_1) H_j(\alpha_2) H_k(\alpha_3) \cdot \exp\{-\alpha_1^2 - \alpha_2^2 - \alpha_3^2\}, \tag{2}$$

where

$$\alpha_i = \frac{u_i - c_i^n}{s_i^n}, \qquad i = 1, 2, 3,$$

the vector $\mathbf{c}^n = (c_1^n, c_2^n, c_3^n)$ is the center of the expansion, and the numbers s_1^n, s_2^n, s_3^n are appropriate scales; $\mathbf{c}^n$ and the s_i^n are allowed to vary with both $\mathbf{x}$ and t, and the summations are to be carried out over $0 \leqq i \leqq L_1$, $0 \leqq j \leqq L_2$, $0 \leqq k \leqq L_3$. The position vector $\mathbf{x}$ will assume a finite number of values in the region of physical interest. Our aim is to evaluate $f(\mathbf{x}, \mathbf{u}, (n+1)\Delta t)$ knowing $f(\mathbf{x}, u, n\,\Delta t)$; $f(\mathbf{x}, \mathbf{u}, (n+1)\Delta t)$ will be evaluated at a set of points

$(\mathbf{x}, \mathbf{u})$ chosen so that a representation of the form (2) can be constructed at the time $t = (n+1)\Delta t$). We first pick new centers $\mathbf{c}^{n+1}$ and new scales s_i^{n+1}, $i = 1, 2, 3$, and evaluate $f(\mathbf{x}, \mathbf{u}, (n+1)\Delta t)$ at points $(\mathbf{x}, \mathbf{u})$, where $\mathbf{u}$ has the components $(c_1^{n+1} + s_1^{n+1}\xi_i, c_2^{n+1} + s_2^{n+1}\xi_j, c_3^{n+1} + s_3^{n+1}\xi_k)$, ξ_i, ξ_j, ξ_k being, respectively, roots of $H_{N_1}(u) = 0, H_{N_2}(u) = 0, H_{N_3}(u) = 0$, N_1, N_2, N_3 to be determined. These values of $f(\mathbf{x}, \mathbf{u}, (n+1)\Delta t)$ can be found by an algorithm of the form

$$f^{n+1}(\mathbf{x}, \mathbf{u}) \equiv f(\mathbf{x}, \mathbf{u}, (n+1)\Delta t) = Af^n(\mathbf{x}, \mathbf{u}) + \Delta t\, Q(f, f), \tag{3}$$

where A is a linear operator such that $f^{n+1} - Af^n$ approximates $\partial f/\partial t - (\mathbf{u} \cdot \nabla) f$, and $Q(f, f)$ is the approximation to the collision integral. In the evaluation of the collision integral, the crucial fact to remember is that f, given by (2), is defined for all arguments $\mathbf{u}$, and thus no further interpolation is required; and in particular Gaussian quadrature can be applied without further ado. Using the representation (2) and obvious changes of variables, the collision term can be reduced to the form

$$C(\mathbf{x}, \mathbf{u}) \int_{-1}^{+1} d\phi \int_{-1}^{+1} d\chi \int_{-\infty}^{+\infty} du_1' \int_{-\infty}^{+\infty} du_2' \int_{-\infty}^{+\infty} du_3'\, G(\phi, \chi, \mathbf{x}, \mathbf{u}') \exp\{-u_1'^2 - u_2'^2 - u_3'^2\}$$

which can be approximated by the weighted Gaussian quadrature formula

$$Q(f, f) = c(\mathbf{x}, \mathbf{u}) \sum_{i=0}^{M_1} \sum_{j=0}^{M_2} \sum_{k=0}^{M_3} \sum_{l=0}^{M_4} \sum_{m=0}^{M_5} G(\phi_i, \chi_j, u_{1,k}', u_{2,l}', u_{3,m}') W_i W_j w_k w_l w_m,$$

where the θ_i, χ_j are roots of appropriate Legendre polynomials, the $u_{1,k}'$, $u_{2,l}'$, $u_{3,m}'$ are roots of appropriate Hermite polynomials, and the W_i, W_j, w_k, w_l, w_m are the corresponding quadrature weights (see [16]).

If $f(\mathbf{x}, \mathbf{u}, (n+1)\Delta t)$ has an expansion of the form (2), with the new centers $\mathbf{c}^{n+1}$ and scales s_i^{n+1}, we have

$$a_{ijk}^{n+1}(\mathbf{x}) = \pi^{-3/2}(s_1 s_2 s_3)^{-1/2} \iiint f(\mathbf{x}, \mathbf{u}, (n+1)\Delta t) H_i(\alpha_1) H_j(\alpha_2) H_k(\alpha_3)\, d\mathbf{u}$$

$$\alpha = (u_i - c_i)/s_i, \quad i = 1, 2, 3,$$

or

$$a_{ijk}^{n+1} = \pi^{-3/2}(s_1 s_2 s_3)^{-1/2} \iiint f(\mathbf{x}, \mathbf{u}, (n+1)\Delta t) H_i(\alpha_1) H_j(\alpha_2) H_k(\alpha_3) \times \exp\{\alpha_1^2 + \alpha_2^2 + \alpha_3^2\} \exp\{-\alpha_1^2 - \alpha_2^2 - \alpha_3^2\}\, d\mathbf{u},$$

which, after an obvious change of variable, can be reduced to a multiple of an integral of the form

$$\iiint g(\alpha_1, \alpha_2, \alpha_3) \exp\{-\alpha_1^2 - \alpha_2^2 - \alpha_3^2\}\, d\alpha_1\, d\alpha_2\, d\alpha_3 .$$

If the a_{ijk}^{n+1} are negligibly small for $i > L_1, j > L_2, k > L_3$, then this last integral is approximately equal to (see [12])

$$\sum_{i=0}^{L_1} \sum_{j=0}^{L_2} \sum_{k=0}^{L_3} g(\alpha_{1,i}, \alpha_{2,j}, \alpha_{3,k}) w_i w_j w_k ,$$

where $\alpha_{1,i}$, $\alpha_{2,j}$, $\alpha_{3,k}$ are roots of $H_{L_1}(u) = 0$, $H_{L_2}(u) = 0$, $H_{L_3}(u) = 0$, respectively, and w_i, w_j, w_k are appropriate quadrature weights. Because of our choice of points at which f^{n+1} is evaluated, g is already known at the required points $\alpha_{1,i}$, $\alpha_{2,j}$, $\alpha_{3,k}$. Once the a_{ijk}^{n+1} have been determined, one can proceed to the next step.

In order to apply the algorithm just described, one needs an initial function f^0, as well as boundary conditions on f^n. The operator A must be such that the scheme

$$f^{n+1} = Af^n$$

is stable. Care must be exercised when the boundary conditions are imposed: $f(\mathbf{x}, \mathbf{u})$ at the boundary may be imposed only for values of $\mathbf{u}$ such that $\mathbf{u}$ points into the gas; the distribution of the velocities of the particles which reach the boundary is determined by the flow and cannot be imposed arbitrarily, (see e.g., [8]). If this obvious condition is violated, instability will result.

In summary, one method can be described as follows: The distribution function f is approximated at each time step by two distinct discrete representations, once as a set of functional values at appropriate points, and once as a set of coefficients in a Hermite expansion. The two representations are related by weighted Gaussian quadrature formulas. The main function of the Hermite representation is to provide a stable and accurate interpolation procedure for use in the evaluation of the collision term. The use of Hermite polynomials is suggested by the known properties of f, in a manner analogous to the choice of weights in Gaussian quadrature. The two representations of f are of equal significance. The use of Hermite series was of course suggested by the work of Grad [7]. It should be noted however that unlike Grad's method, our algorithm allows the number of polynomials used, as well as their scales, to depend on the course of the computation.

The principle of our method was already presented in [4]; unfortunately, the program used and listed in [4] contains programming errors. A variant of the method, using Monte-Carlo rather than Gaussian quadrature for the evaluation of the collision integral, was sketched in [3].

3. Flow in One Space Dimension

We shall now specialize equation (1), as well as our algorithm, to the case of flow in a single space dimension. We assume that f depends only on a single space variable $x = x_1$,

$$f(\mathbf{x}, \mathbf{u}) \equiv f(x, u_1, u_2, u_3),$$

and furthermore, that f is invariant under rotation in the u_2, u_3-plane. We introduce the function

$$F(x, u_1, u_2, t) = f(x, u_1, u_2, 0, t),$$

and the notations $u = u_1$, $u_r = u_2$. Clearly, under our assumptions,

$$F(x, u, u_r, t) = F(x, u, -u_r, t),$$

and for any function ϕ of x, u, and u_r, we have

$$\int_{-\infty}^{+\infty} du_1 \int_{-\infty}^{+\infty} du_2 \int_{-\infty}^{+\infty} du_3\, f(x, u_1, u_2, u_3)\phi(x, u_1, u_2, u_3)$$
$$= \int_{-\infty}^{+\infty} du \int_0^{\infty} du_r \cdot 2\pi u_r \cdot F(x, u, u_r)\phi(x, u, u_r).$$

$F(x, u, u_r)$ satisfies the following equation:

$$\frac{\partial F}{\partial t} + u\frac{\partial F}{\partial x}$$
$$= \sigma^2 \int_0^{\pi} d\phi \sin\phi \int_0^{\pi} d\chi \int_0^{+\infty} du' \int_{-\infty}^{+\infty} du_2' \int_{-\infty}^{+\infty} du_3' \cdot (F^+F^{+\prime} - FF')\,|\mathbf{V}\cdot\mathbf{e}|, \tag{4}$$

where

$$F = F(x, u, u_r)\,,$$

$$F' = F(x, u', u_r')\,,$$

$$F^+ = F(x, u^+, u_r^+)\,,$$

$$F^{+\prime} = F(x, u^{+\prime}, u_r^{+\prime})\,,$$

$$\mathbf{e} = (\cos\phi, \sin\phi\cos\chi, \sin\phi\sin\chi)\,,$$

$$\mathbf{u}' = (u', u_2', u_3')\,, \qquad u_r' = \sqrt{u_2'^2 + u_3'^2}\,,$$

$$\mathbf{u} = (u, u_r, 0)\,,$$

$$\mathbf{V} = \mathbf{u}' - \mathbf{u}\,,$$

$$\mathbf{u}^+ = \mathbf{u} + (\mathbf{V}\cdot\mathbf{e})\mathbf{e} = (u^+, u_2^+, u_3^+)\,,$$

$$\mathbf{u}^{+\prime} = \mathbf{u}' - (\mathbf{V}\cdot\mathbf{e})\mathbf{e} = (u^{+\prime}, u_2^{+\prime}, u_3^{+\prime})\,,$$

$$u_r^+ = \sqrt{(u_2^+)^2 + (u_3^+)^2}\,,$$

$$u_r^{+\prime} = \sqrt{(u_2^{+\prime})^2 + (u_3^{+\prime})^2}\,.$$

The expansion (2) now takes the form

$$F(x, u, u_r, t) = \pi^{-3/2} s_1^{-1/2} s_2^{-1} \sum_{ij} a_{ij}(x, t) H_i(\alpha) H_j(\alpha_r) \exp\{-\alpha^2 - \alpha_r^2\}\,, \tag{5}$$

where

$$\alpha = (u - c^n)/s_1^n\,, \qquad \alpha_r = u_r/s_2^n\,,$$

and

$$0 \leqslant i \leqslant L_1\,, \qquad 0 \leqslant j \leqslant L_2.$$

Given $F(x, u, u_r, n\,\Delta t)$, $F(x, u, u_r, (n+1)\,\Delta t)$ is evaluated at the points $(x_k, u_i, u_{r,j})$, where $u_i = c + s_1 \xi_i$, $u_{r,j} = s_2 \xi_j$, ξ_i, ξ_j being the roots of $H_N(u)$ for an appropriate N. The coefficients $a_{ij}(x, (n+1)\,\Delta t)$ are now given by

$$a_{ij}(x, (n+1)\,\Delta t) = \pi^{-1}(s_1 s_2)^{-1/2} \int_{-\infty}^{+\infty} du \int_{-\infty}^{+\infty} du_r\, F(x, u, u_r) H_i(\alpha) H_j(\alpha_r)\,;$$

a_{ij} can again be evaluated by Gaussian quadrature, the values of F at the appropriate arguments being already known. For j odd, $a_{ij} = 0$.

When F is known in the form (5), the various moments of F can be evaluated using the identities

$$\pi^{-1/2}\int_{-\infty}^{+\infty} u^n\, e^{-u^2}\, du = \begin{cases} 0 & \text{for } n \text{ odd}, \\ (n-1)(n-3)\cdots(1)\cdot 2^{-n/2} & \text{for } n \text{ even}, \end{cases}$$

and

$$\pi^{-1}\int_0^{\infty} (2\pi u_r)\, u_r^n\, e^{-u_r^2}\, du_r = (n)(n-2)\cdots(2)2^{-n/2} \qquad \text{for } n \text{ even}.$$

A natural choice (but by no means the only choice) for c, s_1, and s_2 is the following:

$$c^{n+1} = v \equiv \bar{u}/\rho,$$

$$s_1^{n+1} = \left\{2R\int_{-\infty}^{+\infty} du \int_0^{+\infty} du_r (2\pi u_r)\, u^2\, F(x, u, u_r, n\,\Delta t)\right\}^{1/2},$$

$$s_2^{n+1} = \left\{R\int_{-\infty}^{+\infty} du \int_0^{\infty} du_r (2\pi u_r)\, u_r^2\, F(x, u, u_r, n\,\Delta t)\right\}^{1/2},$$

i.e., the expansion at time $(n+1)\,\Delta t$ has as center the mean velocity at time $n\,\Delta t$, and as scales the directional temperatures at time $n\,\Delta t$. A different choice was made in [4].

Let $x_k = k\,\Delta x$, k being an integer and Δx a special increment. Denote $F(x_k, u_i, u_{r,j}, n\,\Delta t)$ by $F_{i,j,k}^n$. Formula (4) takes the form

$$F_{i,j,k}^{n+1} = AF_{i,j,k}^n + \Delta t\, Q(F, F),$$

where $Q(F, F)$ is again an approximation to the collision term by Gaussian quadrature, and $F^{n+1} - AF^n$ approximates $\partial F/\partial t - u\partial F/\partial x$. A useful choice of A is given by

$$AF_{i,j,k} = \left(1 - u\frac{\Delta t}{\Delta x}\right) F_{i,j,k} + u\frac{\Delta t}{\Delta x} F_{i,j,k+s(u)},$$

where

$$s(u) = \begin{cases} 1 & \text{for} \quad u < 0, \\ -1 & \text{for} \quad u > 0. \end{cases}$$

This A is only of first order accuracy but it has several advantages besides its intuitive appeal; it simplifies coding, and at boundaries ensures automatically that only relevant boundary conditions are used. We shall use this A below.

For stability, we must of course have

$$\frac{\Delta t}{\Delta x} \max_i |u_i| < 1 . \tag{6}$$

The evaluation of $Q(F, F)$ will involve a five-fold summation, i.e., a two-fold summation over roots of Legendre polynomials to handle the two angular integrations, and a three-fold summation over roots of Hermite polynomials to approximate the integrations over u_1', u_2', u_3'.

4. Application to a Shock Problem

Consider a gas of elastic spheres flowing in $-\infty \leqq x \leqq +\infty$, with

$$F(-\infty, u, u_r) = \rho_1 \pi^{-3/2} U_1^{-3} \exp\{-((u - c_1)^2 + u_r^2)/U_1^2\}, \tag{7}$$

$$F(+\infty, u, u_r) = \rho_2 \pi^{-3/2} U_2^{-3} \exp\{-((u - c_2)^2 + u_r^2)/U_2^2\}. \tag{8}$$

By definition, $U_1 = \sqrt{2RT_1}$, $U_2 = \sqrt{2RT_2}$; clearly,

$$\bar{u}(-\infty) = \rho_1 c_1, \qquad \bar{u}(+\infty) = \rho_2 c_2.$$

The Mach number M is defined by

$$M = \sqrt{\frac{6}{5}} \frac{c_1}{U_1}. \tag{9}$$

For $M > 1$ a shock will develop. There is a steady shock if the following conservation laws are satisfied:

$$\rho_1 c_1 = \rho_2 c_2, \tag{10}$$

$$\rho_1(c_1^2 + \tfrac{1}{2}U_1^2) = \rho_2(c_2^2 + \tfrac{1}{2}U_2^2), \tag{11}$$

$$\rho_1 c_1(c_1^2 + \tfrac{5}{2}U_1^2) = \rho_2 c_2(c_2^2 + \tfrac{5}{2}U_1^2), \tag{12}$$

where it is assumed that the ratio of specific heats is

$$\gamma = \tfrac{5}{3}.$$

From equations (10), (11), (12) we may deduce

$$\left(\frac{U_2}{U_1}\right)^2 = \frac{(M^2+3)(5M^2-1)}{16M^2}, \tag{13}$$

$$\frac{c_2}{c_1} = \frac{M^2+3}{4M^2}. \tag{14}$$

An important parameter in the shock problem is the shock thickness, conventionally (and awkwardly) defined by

$$X = \frac{c_2 - c_1}{\max_x |dv/dx|}. \tag{15}$$

The left-hand side of the shock is the upstream side. We pick $\rho_1 = 1, c_1 = 1$. Given M, equations (10), (13), (14) yield u_1, ρ_2, c_2, u_2 which will lead to a steady shock (assuming of course that our frame of reference is moving with the velocity of the shock). We now pick our units so that the upstream mean free path $(\sqrt{2}\pi\rho\,\sigma^2)^{-1}$ is one, i.e., we choose $\sigma^2 = 1/\sqrt{2}\pi$; in these units, X is the ratio of the shock thickness to the mean free path. M completely determines the structure of the shock. The dependence of X^{-1} on M, among other parameters, has been tabulated by a number of authors using various approximate theories.

For practical reasons we replace the region $-\infty \leqq x \leqq +\infty$ by the region $-a \leqq x \leqq a$, where a is chosen large enough so that any further increase will have no visible effect on the shock. At $x = a$ we impose the boundary condition

$$F(a, u, u_r) = \rho_2\,\pi^{-3/2}\,U_2^{-3}\exp\{-((u-c_2)^2 + u_r^2)/U_2^2\} \qquad \text{for} \qquad u < 0\,,$$

and at $x = -a$ we impose the condition

$$F(-a, u, u_r) = \pi^{-3/2}\,U_2^{-3}\exp\{-((u-1)^2 + u_r^2)/U_1^2\} \qquad \text{for} \qquad u > 0\,.$$

We divide $[-a, +a]$ into K segments, with a spatial increment

$$\Delta x = 2a/K\,.$$

Our aim is to obtain the steady solution as a limit, for large time, of a flow which evolves from an initial distribution $F^0 = F(x, u, u_r, 0)$. The choice of F^0, as well as that of the various numerical parameters, will be described below.

The question of numerical stability has not yet been fully discussed. It is clearly necessary that condition (6) be satisfied, and that the boundary conditions be correctly set. These conditions are not sufficient when the number of nonzero coefficients a_{ij} is larger than 4 or 5; the difficulty arises from the infinite range of integration in $\mathbf{u}'$ in the collision integral. This can be seen as follows: the coefficients a_{ij} are always tainted with some numerical error, and since the Hermite polynomials grow when their argument grows, these errors may be dangerously amplified. In fact, expansion (2) is not uniformly valid in $\mathbf{u}$. The resulting instability can be cured by a process we shall call support truncation. Let D be the domain

$$D = \{u, u_r \mid c - s_1 B \leqslant u \leqslant c + s_1 B, \ -s_2 B \leqslant u_r \leqslant s_2 B\} .$$

Let B be large enough so that D contains all the points $(u_i, u_{r,j})$ at which F^{n+1} is evaluated. This implies that if F^{n+1} is evaluated at the roots of $H_N(u) = 0$, then B is larger than all these roots. Let P be the projection operator defined by

$$PF = \begin{cases} F & \text{for} \quad (u, u_r) \quad \text{in} \quad D , \\ 0 & \text{for} \quad (u, u_r) \quad \text{not in} \quad D . \end{cases}$$

Now truncate the range of integration in the collision integral through the approximation $Q(F, F) \simeq Q(PF, PF)$ which assumes that a stabilizing B can be found, large enough for F to be negligible outside the domain D. The validity of this assumption will be verified below.

5. Numerical Results

There remains a considerable number of numerical parameters to be chosen: the number $(L_1 + 1)(L_2 + 1)$ of nonzero terms in the expansion (5), the size $2a$ of the region of integration, the spatial increment Δx, the time step Δt, the number of points at which the collision integral is to be evaluated for each x, the number of quadrature points in each evaluation of that integral and the value of the constant B which defines the projection P. Some obvious consistency conditions must be satisfied: the integrals must be evaluated at a sufficient number of points for the desired a_{ij} to be computable, and the integrations must be carried out with sufficient accuracy so that all the polynomials with non-zero coefficients are taken into consideration. In the calculations below, $L_1 = 4$, $L_2 = 3$, so that, at each point, F is described by 18 parameters, in contrast with, the 5 parameters in the one-dimensional version of Grad's approximation (cf. [7]). The integrals are evaluated at 5×3 points for each x, taking into account the symmetry of F in u_r. Each integration requires 5^5 evaluation of F. $B = 2.2$, i.e., larger than the roots of $H_5(u) = 0$. The choice of Δx, Δt will be specified separately. In the range of Mach numbers M we shall consider, a certain amount of experimentation leads to the conclusion that our choices are adequate.

We first verify that the truncation P has little impact on the result. Since no exact conservation of mass, momentum or energy is incorporated into our scheme, the variation in these quantities under conditions where they should remain constant is an indication of the magnitude of the error. In Table I, we exhibit the variation of these quantities when $M = 1$, with $a = 8$, $\Delta x = 2$, $\Delta t = .55180$; these are the values of a, Δx, Δt we shall use below. Moreover, $F^0 = F(-\infty, u, u_r) = F(+\infty, u, u_r)$. If B were infinite the collision integral would vanish and there would be no variation in any of the quantities listed The errors are largest downstream; the main effect of P is to depress slightly the values of ρ and T.

Table I

Effect of Support Truncation

$\Delta x = 2$, $\Delta t = .5518$, $B = 2.2$

$t/\Delta t$	Mass	Momentum	Energy
1	17.97	17.97	28.72
2	17.92	17.92	28.61
3	17.87	17.87	28.49
4	17.83	17.83	28.37
5	17.75	17.76	28.18
10	17.62	17.66	27.87
15	17.50	17.61	27.61

The task now at hand is to pick an initial function F^0. For reasons which will appear later, we make the choice

$$(16) \qquad F(x, u, u_r, 0) = \begin{cases} F(-\infty, u, u_r, 0) & \text{for} \quad x < 0, \\ F(+\infty, u, u_r, 0) & \text{for} \quad x \geqslant 0. \end{cases}$$

On the finite grid we are using, this implies an initial shock of width Δx and totally unrealistic structure. It is understood that the choice (16) places a heavy strain on our representation of the solution before the smoothing due to the collision term has had time to make itself felt.

In Tables II and III, we display the relaxation of the solution from the data (16), with $2a = 16$, $\Delta x = 2$, $\Delta t = .5518$, and at Mach numbers $M = 2$. In Table II, v, the mean velocity, is tabulated as a function of x for small $t/\Delta t$; this should give a qualitative picture of the behavior of the numerical process. In Table III, the instantaneous values of the reciprocal shock width X^{-1}, the maximum in x of $|\partial v/\partial t|$, the total mass, momentum and energy in the region of integration are shown as functions of t for $M = 2$. The changes in v are mostly due to lateral oscillations of the shock center with little change in structure. The most important characteristics of the relaxation to equilibrium are

Table II

v as a Function of x and t

$\Delta x = 2,\ \Delta t = .5518,\ M = 2$

x	$t/\Delta x = 2$	$t/\Delta t = 3$	$t/\Delta t = 5$
−8	1.000	1.000	1.000
−6	1.000	1.000	1.000
−4	1.000	1.000	.999
−2	.978	.918	.834
0	.464	.495	.513
2	.437	.457	.482
4	.437	.439	.455
6	.437	.438	.439
8	.437	.437	.437

its agonizing slowness and lack of monotonicity. The reason for both can be seen by considering the equation of mass conservation

$$\partial_t \rho + \partial_x(\rho v) = 0 .$$

Momentum must be transported to the boundary to allow a change in ρ, and then transported back to allow convergence to a steady state. This points out the danger in picking initial data close to an approximate solution of the steady equations: the rate of change is even slower, and one must be careful not to decide prematurely that convergence has occurred. Of course, in a systematic study of shock structure for a large set of values of M, such choices would be preferable to (16).

Table III

Relaxation to a Steady Shock

$\Delta x = 2,\ \Delta t = .5518,\ M = 2$

$t/\Delta t$	$\max\lvert\partial v/\partial t\rvert$	X^{-1}	Mass	Momentum	Energy
1	.048	.457	30.76	18.05	35.10
2	.063	.412	30.69	18.08	34.80
3	.046	.375	30.64	18.14	34.55
4	.103	.319	30.59	18.14	34.29
5	.047	.284	30.50	18.19	34.03
6	.097	.244	30.43	18.15	33.84
7	.020	.255	30.24	18.14	33.52
8	.087	.229	30.11	18.00	33.34
9	.016	.240	29.88	17.91	33.03
10	.089	.237	29.72	17.70	32.84
11	.021	.228	29.64	17.54	32.51
12	.082	.253	29.28	17.28	32.31
13	.034	.238	29.02	17.08	31.99
14	.070	.255	28.83	16.81	31.81
15	.046	.254	28.62	16.55	31.62

Table IV

Structure of a Shock

$M = 2,\ t = 4.966$

x	v	ρ	T	H/ρ
−8	1.00	1.00	.30	−1.29
−6	.99	1.00	.30	−1.29
−4	.97	1.01	.32	−1.35
−2	.74	1.18	.46	−1.60
0	.50	1.77	.60	−1.57
2	.50	2.12	.58	−1.46
4	.49	2.27	.57	−1.41
6	.46	2.27	.57	−1.44
8	.43	2.28	.62	−1.56

From the listed values of X^{-1} one may conclude that the steady X^{-1} lies between .237 and .255, in contrast to the value $X^{-1} = .381$ derived from the Navier-Stokes equations (cf. [5]), and in fair agreement with the value $X^{-1} = .235$ obtained by Mott-Smith [13], using his approximate theory. This is also in agreement with the Monte-Carlo determination of Bird [1]. At $M = 1.6$ our computation yields X^{-1} between .164 and .184, compared with Mott-Smith's result $X^{-1} = .164$. At $M = 1.4$ the convergence of the algorithm using the initial data (16) is too slow and we only determined that $X^{-1} \leqslant .190$, which contradicts no previous result.

In Table IV, we exhibit a shock structure. This table should be viewed with the effect of P in mind. The known features of the shock are reproduced: monotonic variation of ρ and v, a small maximum in T (see [18]), a dip in the Boltzmann function H/ρ (which is determined only up to an additive constant).

In Table V, we exhibit the coefficient a_{40} as a function of x and t in order to demonstrate (i) that in the initial stages our representation is inadequate (this could be remedied by a change in the initial data), (ii) that the lower

Table V

a_{40} as a Function of x and t

$\Delta x = 2,\ \Delta t = .5518,\ M = 2$

x	$t/\Delta t = 4$	$t/\Delta t = 9$	$t/\Delta t = 14$
−8	0	0	0
−6	0	0	.03
−4	.12	.07	.21
−2	.26	.13	−.01
0	−.52	−.49	−.37
2	−.16	−.29	−.30
4	−.02	−.12	−.16
6	0	−.03	−.06
8	0	0	0

moments of the solution are seemingly independent of the higher moments, and (iii) that the collision operator has a smoothing effect.

6. Conclusions and Generalizations

One of the main effects of the computations presented in the preceeding section is to lend credence to the results of the Mott-Smith theory in [13]. It also suggests that if it were desired to continue such calculation into the high Mach number region it would be appropriate to replace the expansion (2) by a sum of two expansions, with scales and centers determined by conditions upstream and downstream, respectively. The main difficulty in constructing such a double expansion, which would generalize and systematize the work of Mott-Smith and Ziering, *et al.* [19], lies in finding a systematic way for partitioning a function into the sum of two functions with differing scales; it can however be seen that this difficulty can be overcome if the shock is strong enough.

Furthermore, the number of quadrature points in each evaluation of the collision integral is high enough for Monte-Carlo methods, using the tools for variance reduction described in [3], to be competitive; such transition to a partly Monte-Carlo solution, as sketched in [3], would probably be necessary with present computers if problems in more than one dimension were to be solved. With the NYU CDC6600 computer, one time step costs at present 5 to 10 minutes of machine time. It should be noted that the problem of steady shock formation from the initial data (16) places particularly heavy demands on the accuracy of our algorithm.

Bibliography

[1] Bird, G. A. *Shock wave structure in a rigid sphere gas, Rarefied Gas Dynamics*, Suppl. 3, Vol. I, 1965.

[2] Chapman, S., and Cowling, T. G., *The Mathematical Theory of Non-Uniform Gases*, Cambridge University Press, 1958.

[3] Chorin, A. J., *Hermite expansions in Monte-Carlo computation*, J. Comput. Physics, Vol. 8, 1971, pp. 472–482.

[4] Chorin, A. J., *Numerical Solution of the Boltzmann Equation*, AEC Report NYO-1480-173, New York University, 1971.

[5] Gilbarg, D., and Paolucci, D., *The structure of shock waves in the continuum theory of fluids*, J. Rat. Mech. Anal., Vol. 2, 1953, pp. 617–642.

[6] Grad, H., *On the kinetic theory of rarefied gases*, Comm. Pure Appl. Math., Vol. 2, 1949, pp. 311–407.

[7] Grad, H., *The profile of a steady plane shock wave*, Comm. Pure Appl. Math., Vol. 5, 1952, pp. 257–300.

[8] Grad, H., *Principles of the Kinetic Theory of Gases, Handbuch der Physik*, Vol. XII, Springer-Verlag, 1958.

[9] Haviland, J. K., *Determination of shock-wave thickness by the Monte-Carlo method*, Proc. 3rd Symp Rarefied Gas Dynamics, Academic Press, New York, 1963.

[10] Haviland, J. K., *The solution of two molecular flow problems by the Monte-Carlo method*, Methods in Computational Physics, Vol. 4, 1965.

[11] Kaczmarz, S., and Steinhaus, H., *Theorie der Orthogonalreihen*, Warsaw, 1935.
[12] Lanczos, C., *Applied Analysis*, Prentice Hall, Englewood Cliffs, N.J., 1956.
[13] Mott-Smith, H. M., *The solution of the Boltzmann equation for a shock wave*, Physical Review, Vol. 82, 1951, pp. 885–892.
[14] Nordsieck, A., and Hicks, B. L., *Monte-Carlo evaluation of the Boltzmann collision integral*, Proc. 5th Symp. Rarefield Gas Dynamics, Academic Press, New York, 1967.
[15] Sommerfeld, A., *Thermodynamics and Statistical Mechanics*, Academic Press, New York, 1964.
[16] Stroud, A. M., and Secrest, D., *Gaussian Quadrature Formulas*, Prentice Hall, Englewood Cliffs, N.J., 1966.
[17] Wachman, M., and Hamel, B. B., *A discrete ordinate technique for the nonlinear Boltzmann equation with application to pseudo-shock relaxation*, Proc. 5th Symp. Rarefied Gas Dynamics, Academic Press, New York, 1967.
[18] Yen, S. M., *Temperature overshoots in shock waves*, Phys. Fluids, Vol. 9, 1966, pp. 1417–1418.
[19] Ziering, S., Ek, F., and Koch, P., *Two-fluids models for the structure of neutral shock waves*, Phys. Fluids, Vol. 4, 1961, pp. 975–987.

Received May, 1971.

Numerical study of slightly viscous flow

By ALEXANDRE JOEL CHORIN

Department of Mathematics, University of California, Berkeley

(Received 18 September 1972)

A numerical method for solving the time-dependent Navier–Stokes equations in two space dimensions at high Reynolds number is presented. The crux of the method lies in the numerical simulation of the process of vorticity generation and dispersal, using computer-generated pseudo-random numbers. An application to flow past a circular cylinder is presented.

1. Introduction

The Navier–Stokes equations in two space dimensions can be written in the form

$$\partial_t \xi + (\mathbf{u}\,.\,\nabla)\,\xi = R^{-1}\Delta\xi, \qquad (1a)$$

$$\Delta\psi = -\xi, \qquad (1b)$$

$$u = -\partial_y \psi, \quad v = \partial_x \psi, \qquad (1c)$$

where $\mathbf{u} = (u, v)$ is the velocity vector, $\mathbf{r} = (x, y)$ is the position vector, t is the time, ψ is the stream function, ξ is the vorticity, $\Delta \equiv \nabla^2$ is the Laplace operator and R is the Reynolds number. R is assumed to be so large that finite-difference methods are difficult to apply. Equations (1) are to be solved in a domain D, not necessarily finite, with boundary ∂D, and their solution must satisfy the boundary conditions

$$\mathbf{u} = 0 \quad \text{on} \quad \partial D \qquad (2)$$

and the initial condition

$$\mathbf{u}(x, y, t = 0) \quad \text{given in } D. \qquad (3)$$

Consider in particular the problem of flow past a cylinder of finite cross-section D'. In the vicinity of its boundary ∂D a boundary layer will form, whose thickness will be proportional to $R^{-\frac{1}{2}}$ (see, for example, Schlichting 1960, p. 109). Consider furthermore a finite-difference method whose grid is characterized near the boundary layer and in the wake by a mesh width δ. Since it is presumably necessary that a few mesh points fall within the boundary layer, we find that the condition

$$\delta^2 R = O(1) \qquad (4)$$

must be satisfied. Analysis (Chorin 1969a) suggests the more stringent condition

$$\delta R = O(1). \qquad (5)$$

Conditions similar to (4) and (5) were given by Keller & Takami (1966); they indicate that at Reynolds numbers of practical significance the number of mesh points as well as the amount of computational labour required to obtain a solution would be prohibitive. In practice, insuperable difficulties are encountered at

Reprinted with permission from *Journal of Fluid Mechanics*, Vol. 57, No. 4, 1973.

Reynolds numbers of a few hundred. It is therefore of interest to develop a grid-free numerical method in which the values of the velocity field near a boundary are not all computed but are merely sampled, with computational effort concentrated in regions of greatest interest. We shall now present such a method, which relies on a numerical simulation of the process of vorticity generation and dispersal, using computer-generated pseudo-random numbers. A summary of this method was presented in Chorin (1972).

2. Principle of the method

Consider first the flow of an inviscid fluid (i.e. $R = \infty$). Equations (1) reduce to

$$D\xi/Dt = 0, \quad \Delta\psi = -\xi, \tag{6}$$

where D/Dt denotes a total derivative. One could think of solving equations (6) in the absence of boundaries by partitioning the vorticity ξ into a sum of blobs, i.e. writing

$$\xi = \sum_{j=1}^{N} \xi_j, \tag{7}$$

where the functions ξ_j have small support, i.e. vanish outside a small region (or blob) around a point $\mathbf{r}_j$. ψ will then have the form

$$\psi = \sum_{j=1}^{N} \psi_j, \quad \text{with} \quad \Delta\psi_j = -\xi_j. \tag{8}$$

For $|\mathbf{r}-\mathbf{r}_j|$ large, ψ_j will tend to the form

$$\psi_j \cong \frac{\bar{\xi}_j}{2\pi}\log|\mathbf{r}-\mathbf{r}_j|, \quad \bar{\xi}_j = \int \xi_j\,dx\,dy, \tag{9}$$

where $|\mathbf{r}-\mathbf{r}_j|$ denotes the length of the vector $\mathbf{r}-\mathbf{r}_j$. The expression (9) is the stream function of a point vortex; we are thus assuming that distant blobs affect each other as if they were point vortices of appropriate strength $\bar{\xi}_j$. Neighbouring vortex blobs, however, affect each other's motion unlike neighbouring vortices, in particular, the velocity field should remain bounded, while the velocity field induced by a point vortex becomes unbounded near the vortex (Batchelor 1967, p. 95). If the blobs are small, one can assume that the velocity changes little over their area and, furthermore, that the amount of vorticity they contain, $\bar{\xi}_j$, is small, so that their effect on their immediate neighbours is small. These assumptions have now been justified by Dushane (1973). The gist of the analysis is as follows: Euler's equations are written in integral form, and it is then shown that the right-hand sides of equations (10) below are rectangle-rule approximations to the resulting integrals. From this fact, it is deduced that the error converges to zero with the area of the largest of the supports of the ξ_j. Thus we write

$$\psi = \sum_{j=1}^{N} \bar{\xi}_j \psi^0(\mathbf{r}-\mathbf{r}_j),$$

where $\psi^0(r)$ is a fixed function of r such that

$$\psi^0(r)\begin{cases} \sim (1/2\pi)\log r & \text{for} \quad r \text{ large}, \\ \to 0 & \text{as} \quad r \to 0, \end{cases}$$

and we have
$$\xi = \sum_{j=1}^{N} \bar{\xi}_j \xi_j^0, \quad \xi_j^0 = -\Delta\psi^0(\mathbf{r}-\mathbf{r}_j).$$

The motion of the vortex blobs is then described by

$$\frac{dx_i}{dt} = -\sum_{j\neq i} \bar{\xi}_j \frac{\partial \psi^0}{\partial y} \quad (\mathbf{r}-\mathbf{r}_j) \quad (i = 1, \ldots, N), \tag{10a}$$

$$\frac{dy_i}{dt} = \sum_{j\neq i} \bar{\xi}_j \frac{\partial \psi^0}{\partial x} (\mathbf{r}-\mathbf{r}_j) \quad (i = 1, \ldots, N), \tag{10b}$$

where (x_i, y_i) are the components of $\mathbf{r}_i$. This construction can be summarized as follows: if we consider a collection of vortices having a structure and density such that their density approximates the initial vorticity density, and if their motion is determined by equations (10), then their density will continue to approximate the vorticity density at later times. This statement indicates how a small viscosity can be taken into account through a judicious use of the relationship between diffusion and random walks (see, for example, Einstein 1956, p. 15; Wax 1954, p. 9). Consider the diffusion equation

$$\partial_t \xi = R^{-1}\Delta\xi, \quad \xi = \xi(x, y, t),$$

with initial data $\xi(0) = \xi(x, y, t = 0)$. A solution of this equation using random walks can be obtained as follows. Distribute over the x, y plane points of masses ξ_i and locations $\mathbf{r}_i = (x_i, y_i)$, $i = 1, \ldots, N$, N large, in such a way that the mass density approximates $\xi(0)$. Then move the points according to the laws

$$x_i^{n+1} = x_i^n + \eta_1, \quad y_i^{n+1} = y_i^n + \eta_2, \tag{11a, b}$$

where η_1 and η_2 are Gaussianly distributed random variables with zero mean and variance $2k/R$, k being the time step, and where $x_i^n \equiv x_i(nk)$ and $y_i^n \equiv y_i(nk)$. Then the mean density after n steps (11) will approximate $\xi^n \equiv \xi(nk)$. An algorithm for sampling η_1 and η_2 is readily designed (see, for example, Paley & Wiener 1934, p. 146). Boundaries on which ξ is prescribed are readily handled by maintaining a constant density across them and allowing points from both sides to cross at will. (For analyses, see Einstein (1956) and Wax (1954).)

Now approximate equations (10) by an algorithm of the form

$$x_i^{n+1} = x_i^n + ku^{n,\frac{1}{2}}, \quad y_i^{n+1} = y_i^n + kv^{n,\frac{1}{2}}, \tag{12a, b}$$

where $u^{n,\frac{1}{2}}$ and $v^{n,\frac{1}{2}}$ approximate the right-hand sides of equations (10), k is a time step and $x_i^n \equiv x(nk)$ and $y_i^n \equiv y(nk)$ are as above. Then the vorticity density generated by the motion of the vortices according to the laws

$$x_i^{n+1} = x_i^n + ku^{n,\frac{1}{2}} + \eta_1, \tag{13a}$$

$$y_i^{n+1} = y_i^n + kv^{n,\frac{1}{2}} + \eta_2 \tag{13b}$$

will approximate the solution of equations (1).

Place in the flow an obstacle with boundary ∂D. In the case of an inviscid flow only the normal component of $\mathbf{u}$ can be required to vanish on the boundary. This requirement can be satisfied by adding to the flow induced by the vortices a potential flow with velocity on the boundary so designed that it cancels the normal velocity due to the vortices. This potential flow can be found by solving an integral equation on ∂D (see Kellogg 1929, p. 311) and does not require the imposition of a grid on D. For details, see below. When R is finite, the tangential component of $\mathbf{u}$ has to vanish on ∂D as well. Suppose that at some time t the flow we have so far, which is the sum of the flow due to the vortices and of a potential flow, fails to satisfy this second boundary condition. The effect of viscosity will be to create a thin boundary layer which will ensure a smooth transition from the boundary to the flow inside D. The vorticity in that boundary is readily evaluated; it can then be partitioned among vortex blobs and the latter can be allowed to diffuse according to the laws (13). Once this has occurred, $u^{n,\frac{1}{2}}$ and $v^{n,\frac{1}{2}}$ will be small, and in the neighbourhood of a boundary the random component of equations (13) will be dominant. When a vortex, new or old, crosses ∂D it disappears. This process imitates the physical process of vorticity generation (see the discussion in Batchelor (1967, p. 277)).

It is clear that our method can be applied to flows in finite domains as well as to flows in exterior regions. The example of flow past an obstacle does, however, indicate an advantage of our method: no asymptotic expansion of the solution far from the body need be known in advance.

3. Implementation of the method

We shall now give details of the algorithm just outlined by presenting an explicit form for the blob stream function $\psi^0(r)$ and a construction of $u^{n,\frac{1}{2}}$ and $v^{n,\frac{1}{2}}$ to be used in (13). The method of calculating the potential component of the flow will be presented in the next section.

Consider blob stream functions of the form

$$\psi^0(r) = \begin{cases} (2\pi)^{-1}\log r & (r \geqslant \sigma), \\ (2\pi)^{-1}\, r/\sigma & (r < \sigma), \end{cases} \tag{14}$$

where $r = |\mathbf{r}|$ and σ is a cut-off length, to be determined later. The reason for considering this particular form will appear below. The total circulation around a vortex of this form is 1, and the associated velocity field is continuous and bounded. Assume that at time $t = nk$ we have a velocity field $\mathbf{u}^n$ with vorticity approximated by

$$\xi^n = -\sum_{j=1}^{N} \overline{\xi_j}\,\Delta\psi^0(\mathbf{r}-\mathbf{r}_j) \quad (\Delta \equiv \nabla^2).$$

We now present a sequence of steps which will yield ξ^{n+1}.

Divide the boundary ∂D into M segments of equal length h, with centres Q_i, $i = 1, \ldots, M$; let the co-ordinates of Q_i be (X_i, Y_i). Let $\mathbf{u}_\xi = (u_\xi, v_\xi)$ be the velocity induced by the vortices present at time $t = nk$; we have at $\mathbf{r} = (x, y)$

$$u_\xi(\mathbf{r}) = \frac{1}{2\pi}\sum_j{}_1 \frac{y_j - y}{r_j^2}\overline{\xi_j} + \frac{1}{2\pi}\sum_i{}_2 \frac{y_i - y}{\sigma r_i}\overline{\xi_i}, \tag{15a}$$

$$v_\xi(\mathbf{r}) = -\frac{1}{2\pi}\sum_j{}_1 \frac{x_j - x}{r_j^2}\overline{\xi_j} - \frac{1}{2\pi}\sum_i{}_2 \frac{x_i - x}{\sigma r_i}\overline{\xi_i}, \tag{15b}$$

where $r_j = [(x_j - x)^2 + (y_j - y)^2]^{\frac{1}{2}}$, Σ_1 is a sum over all vortices such that $r_j > \sigma$ and Σ_2 is a sum over all vortices such that $r_i \leqslant \sigma$. Let $\mathbf{n} = (n_1, n_2)$ be the outward normal to ∂D. We find a potential flow $\mathbf{u}_p$ such that $\mathbf{u}_p.\mathbf{n} = -\mathbf{u}_\xi.\mathbf{n}$ (at Q_i, $i = 1, \ldots, M$). The details of the evaluation of $\mathbf{u}_p$ will be presented in the next section. $\mathbf{u}_p + \mathbf{u}_\xi$ satisfies the normal boundary condition on ∂D. We write

$$\mathbf{u}^{n,\frac{1}{2}} = (u^{n,\frac{1}{2}}, v^{n,\frac{1}{2}}) = \mathbf{u}_p + \mathbf{u}_\xi,$$

and use this velocity field in equations (13) to advance the position of the existing vortices; those vortices which cross ∂D are eliminated.

Let $\mathbf{s}$ be a unit vector tangent to ∂D. The total vorticity in the boundary layer which appears when the condition $\mathbf{u}.\mathbf{s} = 0$ is applied is $(\mathbf{u}_p + \mathbf{u}_\xi).\mathbf{s}$ per unit length of ∂D. We now partition the resulting vortex sheet into M blobs, centred at the Q_i. We evaluate $(\mathbf{u}_p + \mathbf{u}_\xi).\mathbf{s}$ at Q_i, and assign to the newly created vortices the vorticity $\bar{\xi} = (\mathbf{u}_p + \mathbf{u}_\xi).\mathbf{s}h$. The newly created vortices cannot be point vortices, since the flow field in the neighbourhood of a point vortex is very different from that near a vortex sheet; in particular, it is not bounded, while in the neighbourhood of a vortex sheet, the velocity does remain bounded, with its tangential components suffering a jump as the sheet is crossed. It is clear that an array of vortices with the structure (14) will approximate these features, since if one draws a line through the centre of such a vortex the velocity field where $r < \sigma$ has a constant magnitude and changes sign abruptly at the centre. Furthermore, as a vortex of this structure leaves the surface, its induced velocity field must exactly annihilate the tangential velocity at the boundary. This condition can be satisfied if

$$\sigma = h/2\pi, \tag{16}$$

and thus the cut-off σ is determined. The newly created vortices then move according to the laws (13); those which leave the fluid disappear; the evaluation of ξ^{n+1} is complete.

The use of equations (15) amounts to an apparently cumbersome method of solution of Poisson's equation. However, the method is intended for use in problems where intense vorticity is confined to small regions, which makes (15) usable, and the alternative methods of solution employ a grid, which would destroy the principle of our method.

4. The evaluation of the potential component of the flow

To complete the description of our algorithm, we now describe how the potential component $\mathbf{u}_p$ is evaluated (see Smith 1970). The equation to be solved is

$$\Delta\psi = -\xi = -\nabla \times \mathbf{u} = 0, \tag{17}$$

subject to the boundary condition

$$\mathbf{u}.\mathbf{n} = -\mathbf{u}_\xi.\mathbf{n} \quad \text{on} \quad \partial D.$$

Equation (17) can be satisfied by a flow of the form

$$\mathbf{u} = \nabla\phi, \tag{18}$$

where ϕ has the form

$$\phi(\mathbf{r}) = \frac{1}{2\pi}\int_{\partial D} \alpha(q) \log R(q)\, dq, \tag{19}$$

where q is a point on ∂D, with co-ordinates (x_q, y_q), and

$$R(q) = [(x-x_q)^2+(y-y_q)^2]^{\frac{1}{2}}.$$

$\alpha(q)$ is a single-layer source (see Kellogg 1929, p. 311), and satisfies the integral equation

$$\alpha(q)-\frac{1}{\pi}\int_{\partial D}\alpha(q')\,\partial_n(\log R(q'))\,dq' = -2\mathbf{u}_\xi.\mathbf{n}, \tag{20}$$

where ∂_n denotes a derivative in the direction of $\mathbf{n}$. We approximate (20) by a system of linear equations. A source of intensity 1 at Q_i induces at the point $Q_j, i \neq j$, a velocity field with components

$$U_1(ij) = -\frac{1}{2\pi}\frac{X_j-X_i}{R_{ij}^2}, \quad U_2(ij) = -\frac{1}{2\pi}\frac{Y_j-Y_i}{R_{ij}^2},$$

$$R_{ij}^2 = (X_j-X_i)^2+(Y_j-Y_i)^2.$$

We approximate $\alpha(q)$ by the M-component vector $\boldsymbol{\alpha} = (\alpha(Q_1), \ldots, \alpha(Q_M))$, which must thus satisfy the matrix equation

$$\mathbf{A}\boldsymbol{\alpha} = \mathbf{b},$$

where the components of $\mathbf{b}$ are the values of $-\mathbf{u}_\xi.\mathbf{n}$ evaluated at the points Q_i, and the matrix $\mathbf{A}$ has the components

$$\begin{cases} a_{ij} = U_1(ij)\,n_1+U_2(ij)\,n_2 & (i \neq j) \\ a_{ii} = \frac{1}{2}h^{-1} & (i = 1, \ldots, N). \end{cases}$$

The discrete form of (18) and (19) is then

$$\mathbf{u}_p(\mathbf{r}) = \sum_{i=1}^{M}\mathbf{u}_p(i),$$

where

$$\mathbf{u}_p(i) = \begin{cases} \dfrac{1}{2\pi}\alpha(Q_i)\dfrac{\mathbf{r}(Q_i)}{r^2(Q_i)} & \text{if } r(Q_i) \geqslant \frac{1}{2}h, \\ \frac{1}{2}h^{-1}\alpha(Q_i)\,\mathbf{n}(Q_i) & \text{if } r(Q_i) < \frac{1}{2}h, \end{cases}$$

where $\mathbf{r}(Q_i)$ is the vector joining Q_i to $\mathbf{r}$, and $r(Q_i) = |\mathbf{r}(Q_i)|$.

5. Heuristic considerations

The crux of our method is the representation of the flow by a randomly placed set of vortices of similar finite structure. This representation was suggested by the author's work on turbulence theory (Chorin 1969*b*, 1970, 1973), and it may be of interest to summarize the relevant considerations.

Much of the theory of turbulence is concerned with the behaviour of the spectrum of the flow at large frequencies (see Batchelor 1960, p. 103). The reason is that this behaviour seems to be independent of the particular flow under consideration. The hope is that an understanding of this behaviour would suggest a way to incorporate these frequencies into a numerical method; finite-difference methods in particular can handle only a bounded range of frequencies.

The high-frequency range of the spectrum is associated with the less smooth part of the flow. The crucial assumption in the author's work is that the loss of smoothness in incompressible flow does not occur uniformly in each flow but is localized in certain regions, much in the same way as the high-frequency components of one-dimensional compressible flow are occasioned by the appearance of shocks. An argument was given to the effect that these rough regions consist of circular vortices; in order to match the observed spectra these circular vortices must have a core of universal structure. Their locations may be thought of as random. By constructing the flow from such elements, one ensures that the high-frequency range is taken into account. It is interesting to note that similar considerations can be applied to Glimm's (1965) solution of nonlinear hyperbolic systems.

The question now arises as to what is the order of magnitude of the errors induced by our approximation. In the case of infinite R (inviscid flow), we have a fully deterministic method of solving Euler's equations, and as long as k is of order h, we expect the overall error to be of $O(k)$. In the case of finite R, the velocity at any one point or at any one instant is a random variable, and convergence can be expected only in the mean, i.e. as one averages over large regions, or over long times or over ensembles. In the mean, diffusion is represented without error; the crucial problem is to assess the effect of the interaction between the random and deterministic parts of equations (13). As far as the determination of inertial effects is concerned, the random variables η_1 and η_2 can be viewed as harmful perturbations. The standard deviation of η_1 and η_2 is $[2k/R]^{\frac{1}{2}}$. After n steps, the total effect of the random perturbations will be to induce a displacement of order

$$[n.2k/R]^{\frac{1}{2}} = O(R^{-\frac{1}{2}})$$

in the location of the vortices. If this can be identified with an error in the evaluation of the nonlinear terms, its magnitude will be of $O(R^{-\frac{1}{2}})$. We thus conjecture that the mean error in our calculation is of $O(k) + O(R^{-\frac{1}{2}})$. The second term in this estimate may appear shocking, since it does not depend on k. However, the relations (4) and (5) indicate that, if $R^{-\frac{1}{2}}$ is not small, a difference method may be used and our algorithm becomes unnecessary. We therefore do not expect valid solutions at low R.

At the other extreme, some difficulty may be expected at very high Reynolds numbers. This is so because the boundary layers formed by the algorithm are made up of a few bouncing vortices and are thus noisy; turbulence effects should therefore appear at too small a value of R, as they do, for example, in noisy wind tunnels or around rough bodies.

6. Application: flow past a circular cylinder

Consider a circular cylinder of radius 1, immersed in a fluid of density 1, to which is imparted at $t = 0$ a constant velocity of magnitude 1. The Reynolds number based on cylinder radius is $R = \nu^{-1}$, where ν is the viscosity. (In the literature one encounters a Reynolds number $R' = 2R$ based on cylinder diameter.) Let the origin O be fixed in the centre of the cylinder's base, with the negative-x axis pointing in the direction of the motion. In the resulting frame of reference the velocity at infinity is $(1, 0)$, and the cylinder is at rest. ∂D is the circumference of the base.

Divide ∂D into M pieces of length $h = 2\pi/M$. The cut-off length is $\sigma = h/2\pi = 1/M$. One of the important functionals of the flow is the drag coefficient C_D, which in our units is simply the force per unit length of the cylinder. We have

$$C_D = C_\nu + C_p,$$

where C_ν is the skin drag, given by

$$C_\nu = -\frac{1}{R}\int_{\partial D} \xi_\partial \sin\theta \, d\theta, \tag{21}$$

and C_p is the form drag, given by

$$C_p = \int_{\partial D} p_\partial \cos\theta \, d\theta, \tag{22}$$

where $r\cos\theta = x$, $r\sin\theta = y$, ξ_∂ is the vorticity on ∂D and p_∂ is the pressure on ∂D. p_∂ can be evaluated using the formula

$$p_\partial(\theta) = -\frac{1}{R}\int_{\theta_0}^{\theta} \partial_n \xi \, ds + \text{constant}, \tag{23}$$

where $\partial_n \xi$ is the normal derivative of ξ and the integration is carried out along ∂D. The problem at hand is to evaluate ξ and $\partial_n \xi$ given our random array of vortices. Introduce the regions A_j and A_j^+ defined by

$$A_j = \{x, y \mid 1 \leqslant r < 1+\mu, (j-\tfrac{1}{2})\,2\pi/M \leqslant \theta < (j+\tfrac{1}{2})\,2\pi/M\},$$
$$A_j^+ = \{x, y \mid 1+\mu \leqslant r < 1+2\mu, (j-\tfrac{1}{2})\,2\pi/M \leqslant \theta < (j+\tfrac{1}{2})\,2\pi/M\},$$

where $\mu = (2k/R)^{\frac{1}{2}}$ is the standard deviation of η_1 and η_2. $\xi(A_j)$ and $\xi(A_j^+)$ are defined as the sums of the vorticities $\bar{\xi}_i$ associated with vortices whose centres fall within A_j and A_j^+ divided by the areas of A_j and A_j^+ respectively. We now identify $\xi(A_j)$ with $\xi(Q_j)$, and $(\xi(A_j^+)-\xi(A_j))/\mu$ with $\partial_n \xi(Q_j)$. It is worth emphasizing that the grid just introduced is used not to advance the calculation, but only to diagnose its outcome. $\xi(A_j)$ and $\xi(A_j^+)$ are random variables, and can be expected to have substantial variance; we therefore introduce the averaged drag $C_D(t, T)$, defined by

$$C_D(t, T) = \int_{t-T}^{t} C_D(t)\, dt, \tag{24}$$

where $C_D(t)$ is the drag C_D at time t. The integrals (21)–(24), can be evaluated through the use of the trapezoidal rule.

The time step k and the number M of vortices created per time step remain to be chosen. As k decreases M must increase; this is so because the deterministic component of the right-hand sides in equations (13) is proportional to k, while the random component has standard deviation proportional to $\sqrt{k}$. Thus as k is decreased, each vortex has an increasing number of opportunities to cross ∂D and disappear; since a minimum number of vortices must be maintained in the fluid, more and more must be created. For a given k, M is chosen so large that a further increase does not affect the solution. k must be chosen so that a decrease in k will not affect the flow; the solution is rather insensitive to k. After some experimentation the value $k = 0{\cdot}2$ was picked. The required M is 20. All the calculations below were made with these parameters.

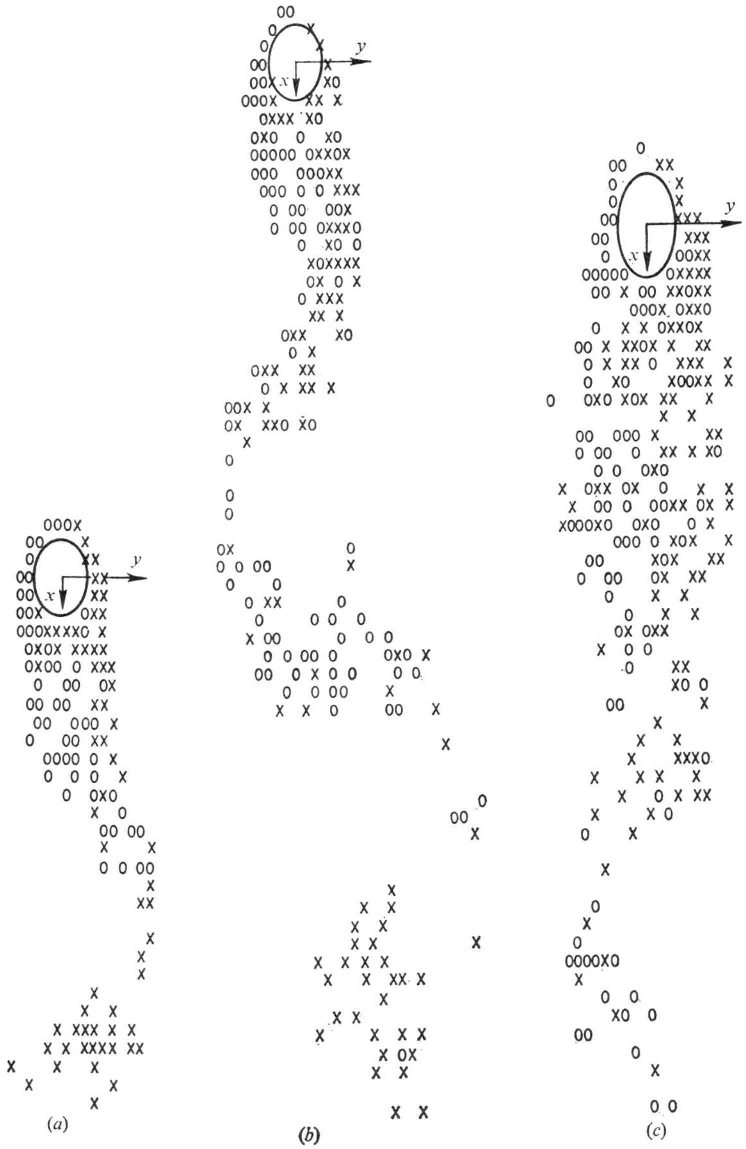

FIGURE 1. Flow at (*a*) $R = 1000$, $t = 12$; (*b*) $R = 1000$, $t = 24$; (*c*) $R = 100$, $t = 16$.

For figures 1(*a*) and (*b*) the flow at $R = 1000$ at two times is visualized. The domain is divided into squares of side $\epsilon = 0{\cdot}3$; if a square contains no vortices nothing is printed; if the sum of the vortices in the square is positive a cross is printed; if the sum is negative a circle is printed. Note the deformation of the circle by the computer printer. This visualization may be crude, but it is in

t	$C_D(t, t)$	$C_D(t, 2)$
2	0·993	0·993
4	1·118	1·232
6	1·020	0·833
8	1·056	1·158
10	1·068	1·118
12	1·034	0·863
16	1·014	0·894
20	1·049	1·216
24	1·060	1·073

TABLE 1. Average drag as a function of time at $R = 1000$

j	$\xi(A_j)$	$\xi(A_j^+)$
1	10·8	14·4
2	6·0	5·9
3	−21·3	−20·9
4	8·1	7·9
5	−25·4	−24·9
6	7·2	7·0
7	52·1	51·1
8	22·7	49·3
9	35·7	35·0
10	20·6	20·2
11	−9·5	−6·4
12	−30·9	−30·3
13	−52·6	−51·6
14	−24·8	−41·9
15	−19·4	−49·3
16	−9·1	−8·9
17	−8·3	−11·4
18	2·3	2·2
19	11·6	11·4
20	5·1	10·0

TABLE 2. Vorticity distribution at $R = 1000$, $t = 10$

keeping with the spirit of our method, in which no location is certain; the visualization is of course most inadequate at boundaries. In table 1 the values of $\xi(A_j)$ and $\xi(A_j^+)$ for $t = 10$ and $R = 1000$ are tabulated. Separation can be detected when $\xi(A_j)$ and $\xi(A_j^+)$ differ appreciably. It can be seen that the separation of the boundary layers occurs (asymmetrically) around $\theta = 126°$ and $\theta = 288°$. The values of $C_D(t,t)$ and $C_D(t,2)$ at $R = 1000$ are tabulated in table 2. The numbers printed yield a mean drag of 1·04, in excellent agreement with experiment (Schlichting 1960, p. 16).

We now decrease R. At $R = 500$ we obtain a mean drag $C_D = 1{\cdot}15$. In figure 1 (*c*) we visualize the flow at $R = 100$, which is at the lower limit of applicability of the method; $\epsilon = 0{\cdot}25$. $C_D(t,2)$ climbs to a maximum of 2·80, and then oscillates between 1·30 and 1·18. (The experimental values lie between 1·20 and 1·25.) The skin drag is $0{\cdot}26 \pm 0{\cdot}02$; the experimental value is 0·28.

As we increase R we find that $C_D = 1{\cdot}09$ at $R = 5000$; this rise is of course experimentally observed. At $R = 10\,000$, C_D is approximately 0·87, about $\frac{2}{3}$ of the experimental value. We can conjecture that the rough representation of the boundary layer triggers a premature onset of the drag crisis, analogous to the effect of a rough boundary or a noisy flow. This conjecture is apparently confirmed by the fact that, at $R = 100\,000$, $C_D = 0{\cdot}29$, in good agreement with the experimental value beyond the drag crisis. However, more thought is required before we are ready to claim that the method is able to follow a transition to turbulence. Beyond $R = 10\,000$ the vortex street behind the cylinder becomes disorderly at about 10 units of length behind the cylinder. In all our calculations, the number of vortices in the fluid at $t = 30$ is approximately 300, and it takes about 12 minutes of CDC 6400 time to follow the evolution from $t = 0$ to $t = 30$.†

7. Conclusion and further work

We have presented a numerical method containing a random element which makes possible the analysis of flow at high Reynolds number with comparatively little effort. The price paid for this achievement is the loss of pointwise convergence in either space or time. The method will be applied to other problems besides the one presented here, but the most fascinating subject for further research, both theoretical and numerical, is the possibility that this method is able to simulate the transition to turbulence.

Another problem under investigation is the development of a similar method for three-dimensional flow problems, in which vortices will be replaced by vortex lines.

This work was carried out while the author was Visiting Miller Research Professor at the University of California, Berkeley, with partial support from the Office of Naval Research under Contract no. N00014-69-A-0200-1052.

REFERENCES

Batchelor, G. K. 1960 *The Theory of Homogeneous Turbulence*. Cambridge University Press.

Batchelor, G. K. 1967 *An Introduction to Fluid Mechanics*. Cambridge University Press.

Chorin, A. J. 1969*a* On the convergence of discrete approximations to the Navier–Stokes equations. *Math. Comp.* **23**, 341.

Chorin, A. J. 1969*b* Inertial range flow and turbulence cascades. *A.E.C. R. & D. Rep.* NYO-1480-135.

Chorin, A. J. 1970 Computational aspects of the turbulence problem. In *Proc. 2nd Int. Conf. on Numerical Methods in Fluid Mech.*, p. 285. Springer.

Chorin, A. J. 1972 A vortex method for the study of rapid flow. In *Proc. 3rd Int. Conf. on Numerical Methods in Fluid Mech.* Springer.

Chorin, A. J. 1973 Representations of random flow. (To appear.)

Dushane, T. E. 1973 Convergence of a vortex method for Euler's equations. (To appear.)

Einstein, A. 1956 *Investigation on the Theory of Brownian Motion*. Dover.

† The computer program used to obtain the results above is available from the author.

GLIMM, J. 1965 Solutions in the large for nonlinear hyperbolic systems of conservation laws. *Comm. Pure Appl. Math.* **18**, 697.

KELLER, H. B. & TAKAMI, H. 1966 Numerical studies of steady viscous flow about cylinders. In *Numerical Solutions of Nonlinear Differential Equations* (ed. D. Greenspan), p. 115. Wiley.

KELLOGG, O. D. 1929 *Foundations of Potential Theory*. New York: Ungar. (Also available through Dover Publications.)

PALEY, R. E. A. C. & WIENER, N. 1934 *Fourier Transforms in the Complex Domain, Colloquium Publications*, vol. 19. Providence: Am. Math. Soc.

SCHLICHTING, H. 1960 *Boundary Layer Theory*. McGraw-Hill.

SMITH, A. M. O. 1970 Recent progress in the calculation of potential flows. *Proc. 7th Symp. Naval Hydrodynamics*. Office of Naval Research.

WAX, N. 1954 *Selected Papers on Noise and Stochastic Processes*. Dover.

Discretization of a Vortex Sheet, with an Example of Roll-Up*

ALEXANDRE JOEL CHORIN†

Department of Mathematics, University of California, Berkeley 94720

AND

PETER S. BERNARD‡

Department of Mechanical Engineering, University of California, Berkeley 94720

Received May 9, 1973

The point vortex approximation of a vortex sheet in two space dimensions is examined and a remedy for some of its shortcomings is suggested. The approximation is then applied to the study of the roll-up of a vortex sheet induced by an elliptically loaded wing.

INTRODUCTION

The study of the motion of vortex sheets in two-dimensional space is of great importance in a number of practical problems [1, 2, 5], as well as in the design of numerical algorithms [3]. Rosenhead [10] introduced a method of analysis in which the sheet is approximated by an array of point vortices; this method was applied by Westwater [12] to the roll-up problem. Recently, Takami [11] and Moore [8] have shown that this method can produce errors of arbitrarily large magnitude, and thus cast a doubt on its validity. On the other hand, other vortex methods, involving interpolation [4] or a cut-off [3], have recently achieved notable successes in different contexts; one of them has even been proved to be convergent [6]. This discrepancy is of substantial interest, and it is the purpose of the present paper to contribute to its resolution.

It is fairly obvious that a point vortex approximation to a vortex sheet (or for

* Partially supported by the Office of Naval Research under Contract No. N00014-69-A-0200-1052.

† Alfred P. Sloan Research Fellow.

†† National Science Foundation graduate Fellow

Reprinted with permission from *Journal of Computational Physics*, Vol. 13, No. 3, 1973.

that matter, to a continuous vorticity distribution) cannot be taken too literally, since a point vortex induces a velocity field which becomes unbounded, and cannot approximate a bounded field in any reasonable norm. Thus the results of Takami and Moore are understandable. However, we conjecture that as soon as the velocity field of the point vortices is smoothed out and made bounded, i.e., the point character of the point vortices is not taken too literally, the approximation becomes reasonable. Such smoothing occurs in all successful applications, and in fact one may conjecture that the old results of Westwater are better than the new results of Takami and Moore because the limited accuracy of precomputer calculation had a smoothing effect. We intend to present numerical evidence in support of this conjecture in an application to a problem involving the roll-up of a vortex sheet.

Approximation of a Vortex Sheet by a Finite Array of Point Vortices

Consider a vortex sheet whose vorticity is parallel to its length and whose two-dimensional cross section initially lies on the curve C: $x = x(s)$, $y = y(s)$, where x and y are the coordinates of a point and s is the arc length. The vorticity distribution is $\xi = \xi(s)$. The sheet will be moved by the velocity field which it induces; the equations of motion are

$$\frac{\partial x}{\partial t} = -\partial_y \int_C \frac{1}{2\pi} \log | \mathbf{r}(s) - \mathbf{r}(s')| \, \xi(s')\, ds' \tag{1a}$$

$$\frac{\partial y}{\partial t} = \partial_x \int_C \frac{1}{2\pi} \log | \mathbf{r}(s) - \mathbf{r}(s')| \, \xi(s')\, ds' \tag{1b}$$

where $\mathbf{r} = (x, y)$, $| \mathbf{r} | = | x^2 + y^2 |^{1/2}$, $x = x(t, s)$, $y = y(t, s)$. In Rosenhead's approximation the sheet is replaced by an array of N point vortices, located at the points $\mathbf{r}_i = (x_i, y_i)$ of C, with vorticities ξ_i, $i = 1,..., N$ whose distribution approximates $\xi(s)$. The motion of these N vorticies is then described by the N ordinary differential equations,

$$\frac{dx_i}{dt} = -\partial_y \sum_{j\neq i} \psi(\mathbf{r} - \mathbf{r}_j)\, \xi_j\,, \tag{2a}$$

$$\frac{dy_i}{dt} = \partial_x \sum_{j\neq i} \psi(\mathbf{r} - \mathbf{r}_j)\, \xi_j\,, \tag{2b}$$

where $\psi = (1/2\pi) \log | \mathbf{r} |$ is the stream function associated with a point vortex (see [1]).

It is readily seen that the right-hand sides of Eqs. (2a), (2b) are rectangle rule approximations to the integrals on the right-hand sides of Eqs. (1), and as such can be expected to approximate the latter well, as long as the derivatives of the integrands are not too large, i.e., as long as $|d\xi/ds|$, $|\partial x/\partial s|$, $|\partial y/\partial s|$ are bounded by some constant K which is small compared to h^{-1}, where h is a typical distance between vortices. When this condition is not satisfied, the approximate balance between the flow due to the vortices on either side of a given vortex will no longer hold because of the unduly large flow produced by a point vortex, and the several vortices will capture each other, i.e., start following complicated paths around each other. Once this process starts at any point, it rapidly spreads throughout the sheet; this type of breakdown was well documented in [2, 8, 11]; see also below.

In summary, as soon as the approximation (2) ceases to be accurate because the numerical parameter h becomes comparable with the intrinsic characteristic lengths of the problem, it also becomes qualitatively unreasonable. This breakdown is analogous to the effect of nonlinear instability in a difference scheme (see [9]). In the case of difference schemes, cures are known: One can sometimes reduce h (at the cost of added computational labor) or, if the region of inaccuracy is initially small, one can introduce an artificial viscosity so designed that its effect is local. The analog of this latter technique is smoothing, which we shall now explain on an example.

Vortex Sheet Induced by an Elliptically Loaded Wing

Consider in particular the vortex sheet initially located on the strip $-a \leqslant x \leqslant a$, with vorticity distribution

$$\xi = 2Ux(a^2 - x^2)^{-1/2}.$$

The significance of this particular configuration is explained in [1]. Clearly the

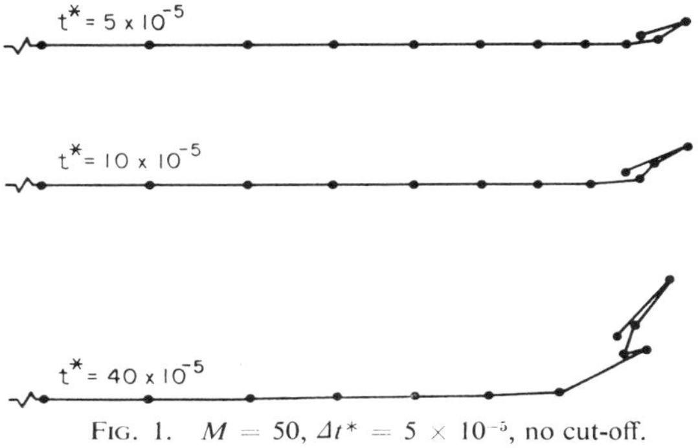

FIG. 1. $M = 50$, $\Delta t^* = 5 \times 10^{-5}$, no cut-off.

vortex point method is likely to fail at the tips of the sheet $x = \pm a$, since at these points $d\xi/dx$ becomes unbounded. A typical breakdown is illustrated in Fig. 1 (which is analogous to the results displayed by Takami [11] and Moore [8]). The strip is divided into $2M$ pieces of equal total vorticity, and a point vortex of strength $\pm 2Ua/M$ is placed at the vorticity centroid of each piece. A dimensionless time $t^* = Ut/a$ is introduced. In the calculation which leads to Fig. 1 we used $M = 50$; equations (2) were integrated using a fourth-order Runge–Kutta method with $\Delta t^* = 5 \times 10^{-5}$. The positions of the last 12 vortices at the extreme right-hand side of the sheet are displayed at times $t^* = 5 \times 10^{-5}$, 10^{-4}, 4×10^{-4}; they are connected in the order in which they were initially placed. Chaos is generated at the tip before the other vortices have had time to move appreciably. The approximation (2) seems to be inapplicable.

Now replace $\psi = (1/2\pi) \log |\mathbf{r}|$ in Eqs. (2) by

$$\psi_\sigma = \begin{cases} (1/2\pi) \log |\mathbf{r}|, & r \geqslant \sigma; \\ (1/2\pi)(|\mathbf{r}|/\sigma) + \text{const} & r < \sigma. \end{cases}$$

σ is a (small) cut-off. The introduction of such a cut-off is analogous to the introduction of a small viscosity which allows the vorticity in a point vortex to diffuse (see [1]). It is artificial rather than real viscosity since its effect is not cumulative; the vorticity is diffused a little, and then spreads no more. If σ is of order a/M^2, only the motion of the few vortices near the tip will be affected. The particular form of ψ_σ was chosen by analogy with the form used in Chorin [3], but in fact almost any form which keeps the velocity field bounded for small enough r seems to be adequate. We again use a Runge–Kutta algorithm, with $\Delta t^* = 5 \times 10^{-5}$, and $\sigma = 6a/M^2$. The results are displayed in Fig. 2 for the times indicated. The spurious

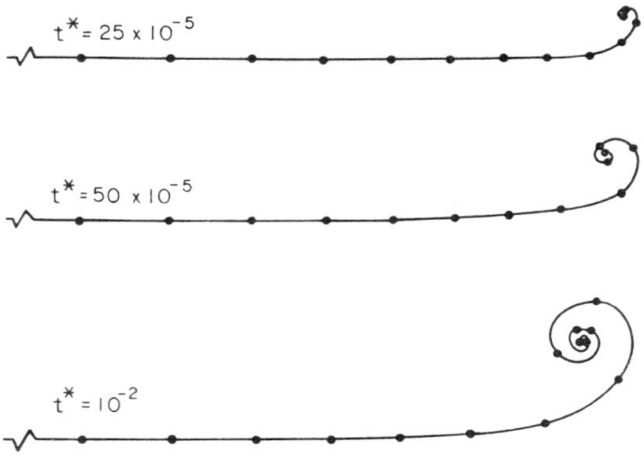

FIG. 2. $M = 50$, $\Delta t^* = 5 \times 10^{-5}$, $\sigma = 6a/M^2$.

motion near the tip has been damped, and the roll-up is proceeding as expected. These results are independent of $\sigma/(a/M^2)$ within a wide range. It is important to note that the introduction of the cut-off does not affect the mutual interaction of distant portions of the sheet; furthermore, the results are not sensitive to the exact choice of ψ_σ for $r < \sigma$.

In choosing a Runge–Kutta integration formula, unevenly spaced vortices and a very small time step, we have followed the example of Takami and Moore, whose aim was to display the breakdown in the approximation. Our aim, of course, was to show that on the contrary, with $\sigma \neq 0$ the roll-up proceeds smoothly under otherwise comparable conditions. If on the other hand, one is interested in the evolution of the sheet, one might as well pick a substantially larger Δt^* (subject to the obvious convergence requirement $U\Delta t \ll a$); it is furthermore reasonable to use a straightforward Euler scheme, which is more economical in terms of computing time and does not introduce errors of larger order than those originating from the replacement of (1) by (2); finally, there is no obvious reason not to use equidistant vortices whose intensity is proportional to the value of ξ at their location. Figures 3 and 4 were obtained under such conditions, with $\Delta t^* = 10^{-2}$. Figure 3 displays the configuration at $t^* = 1$, with $M = 50$, $\sigma = a/M$. These changes effect a saving in computer time, but do not affect the nature of the solution. An effort was made to verify Kaden's result [7] to the effect that the sheet has the form of a spiral with polar equation $r = C(\theta - \theta_0)^{2/3}$ as $\theta \to \infty$. C and θ_0 were determined using the locations of the vortices marked A and B. Vortices 1 through 14 are seen to lie on the resulting spiral with no visible error. Kaden's result has been derived with the help of simplifying assumptions, and is not expected to hold uniformly throughout the spiral [5]. The agreement we obtained is therefore quite satisfactory.

Figure 4 displays the configuration at $t^* = 6.5$, with $\sigma = a/M$. At this time, it is no longer clear that we have an approximation to a sheet. We may as well consider that we have real but small viscosity; the distribution of vorticity in the region of concentrated vorticity is then continuous, and our scheme with cut-off is then

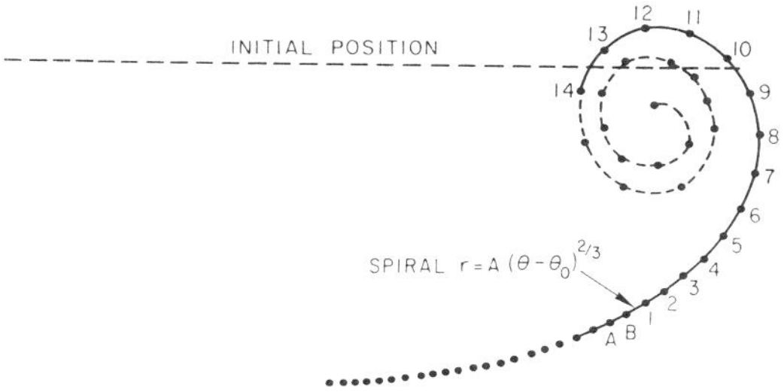

FIG. 3. $\Delta t^* = 10^{-2}$, $M = 50$, $\sigma = a/M$, $t^* = 1$.

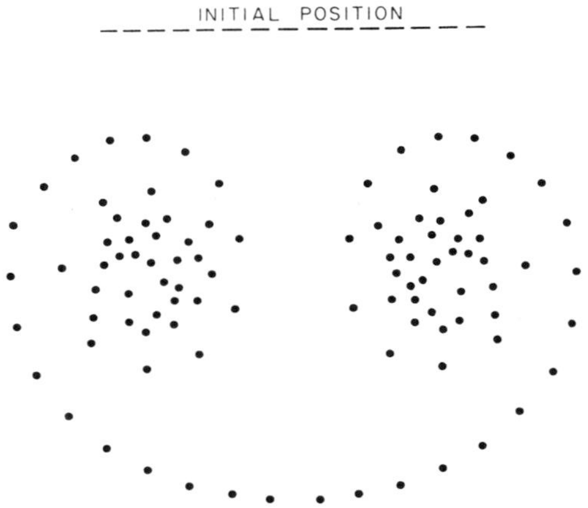

FIG. 4. $\Delta t^* = 10^{-2}$, $M = 50$, $\sigma = 2a/M$, $t^* = 6.5$

known to be valid (see [3, 6]). The results in Figs. 3 and 4 are in fact independent of σ. They are reproduced with only minor differences even when $\sigma = 0$, provided we keep $\Delta t^* = 10^{-2}$; this is consistent with the fact that nonlinear instability has a smaller impact when the time step is increased (see [9]). Thus, one seems to be able to use numerical error as a smoothing mechanism which keeps the point vortex approximation under control.

CONCLUSION

We have shown that Rosenhead's point vortex approximation is valid and useful, provided the singular character of all the vortices is obviated by some smoothing. This fact explains the discrepancy between the failure of some point vortex approximations and the success of others.

REFERENCES

1. A. K. BATCHELOR, "Introduction to Fluid Dynamics," Cambridge University Press, London, 1967.
2. G. BIRKHOFF, Helmholtz and Taylor instability, in "Proc. Symp. Appl. Math.," Vol. 13, Am. Math. Soc., Providence, R.I., 1962.
3. A. J. CHORIN, *J. Fluid Mech.* **57** (1973), 785.
4. J. P. CHRISTIANSEN, Numerical simulation of hydrodynamics by the method of point vortices, submitted to *J. Computational Phys.*

5. W. F. DURAND, "Aerodynamic Theory," Dover, New York, 1934.
6. T. E. DUSHANE, Convergence of a vortex method for Euler's equation, to appear in *Math. Comp.*
7. H. KADEN, *Ingenieur-Archiv. Berlin* **2** (1931), 140.
8. D. W. MOORE, "The Discrete Vortex Approximation of a Finite Vortex Sheet," Calif. Inst. of Tech. Report AFOSR-1804-69, 1971.
9. R. D. RICHTMYER AND K. W. MORTON, "Finite Difference Methods for Initial Value Problems," Interscience, New York, 1967.
10. L. ROSENHEAD, *Proc. Roy. Soc. London, Ser. A* **134** (1932), 170.
11. H. TAKAMI, "Numerical Experiment with Discrete Vortex Approximation, with Reference to the Rolling Up of a Vortex Sheet," Dept. of Aero. and Astro., Stanford Univ., Report SUDAER 202, 1964.
12. F. L. WESTWATER, ARC Report and Memo No. 1692, 1936.

Random Choice Solution of Hyperbolic Systems*

ALEXANDRE JOEL CHORIN

Department of Mathematics, University of California, Berkeley, California 94720

Received April 9, 1976

A random choice method for solving nonlinear hyperbolic systems of conservation laws is presented. The method is rooted in Glimm's constructive proof that such systems have solutions. The solution is advanced in time by a sequence of operations which includes the solution of Riemann problems and a sampling procedure. The method can describe a complex pattern of shock wave and slip line interactions without introducing numerical viscosity and without a special handling of discontinuities. Examples are given of applications to one- and two-dimensional gas flow problems.

OUTLINE OF GOAL AND METHOD

The goal of the present paper is to present a numerical method for solving non-linear hyperbolic systems, in particular those which arise in gas dynamics. The method is meant to be usable when the solution sought exhibits a complex pattern of shock waves and slip lines, and possibly large energy densities; it is meant to be useful when methods which rely on either an artificial viscosity or on a special treatment of discontinuitites become impractical because they are either too difficult to program or too expensive to run. The particular applications we have in mind are to combustion problems in engines, where a complex wave pattern is known to exist [11], and where the use of finite differences is undesirable because numerical viscosity cannot be allowed and because there arise stability problems which are hard to overcome [3].

The method of computation evolved from Glimm's constructive existence proof [5], which will now be described. Consider the strictly nonlinear system of equations

$$\mathbf{u}_t = (\mathbf{f}(\mathbf{u}))_x , \tag{1}$$

where $\mathbf{u}$ is a solution vector and the subscripts denote differentiation. (For a definition of strict nonlinearity, see [8].) Let the initial data $\mathbf{u}(x, 0)$ be close to constant (the specific assumptions are spelled out in [5]). The time t is divided into

* Partially supported by the Office of Naval Research under Contract N00014-69-A-0200-1052.

Reprinted with permission from *Journal of Computational Physics*, Vol. 22, No. 4, 1976.

intervals of length k. Let h be a spatial increment. The solution is to be evaluated at time $t = nk$, n integer, at the points ih, $i = 0, \pm 1, \ldots$, and at time $(n + \frac{1}{2})\,k$ at the points $(i + \frac{1}{2})\,h$. Let $\mathbf{u}_i^n$ approximate $\mathbf{u}(ih, nk)$, and let $\mathbf{u}_{i+1/2}^{n+1/2}$ approximate $\mathbf{u}((i + \frac{1}{2})\,h, (n + \frac{1}{2})\,k)$. To find $\mathbf{u}_{i+1/2}^{n+1/2}$ given $\mathbf{u}_i^n$, $\mathbf{u}_{i+1}^n$ (and thus define the algorithm) one begins by considering an initial value problem for Eqs. (1) with the discontinuous initial data

$$\mathbf{u}(x, 0) = \mathbf{u}_{i+1}^n \qquad \text{for} \quad x \geqslant 0,$$

$$\mathbf{u}(x, 0) = \mathbf{u}_i^n \qquad \text{for} \quad x < 0.$$

(Such a problem is called a Riemann problem.) Let $\mathbf{v}(x, t)$ denote the solution of this Riemann problem, and let θ_i be a value of a random variable θ equidistributed in $[-\frac{1}{2}, \frac{1}{2}]$; let P_i be the point $(\theta_i h, k/2)$, and let $\bar{\mathbf{u}} = \mathbf{v}(P_i) = \mathbf{v}(\theta_i h, k/2)$ be the value of the solution of the Riemann problem at P_i. Set

$$\mathbf{u}_{i+1/2}^{n+1/2} = \bar{\mathbf{u}}.$$

A similar construction allows one to proceed from $\mathbf{u}_{i+1/2}^{n+1/2}$ to $\mathbf{u}_i^{n+1}$. In [5], Glimm showed that under certain conditions, $\mathbf{u}$ converges to a weak solution of (1). In the next four sections, it will be shown how this construction can be made into an efficient numerical tool.

In the first of these sections, the algorithm is applied to the solution of the equations of gas dynamics in one space dimension. The treatment of boundary conditions, the transport of passive quantities, and, most importantly, the sampling procedure for θ, are described. An appropriate choice of sampling procedure is crucial to the success of the method.

In the second of these sections, the method of solution of the Riemann problems is described, following in the main the ideas of Godunov [6].

In the third section, the method is generalized to multidimensional problems. A section is then devoted to examples, and conclusions are given at the end.

An earlier attempt to program Glimm's method is described in [10]. An interesting discussion is given in [9].

Implementation of Glimm's Method

The equations of gas dynamics in one space dimension can be written in the (conservation) form

$$\rho_t + (\rho u)_x = 0, \tag{2a}$$

$$(\rho u)_t + (\rho u^2 + p)_x = 0, \tag{2b}$$

$$e_t + ((e + p)\,u)_x = 0, \tag{2c}$$

where the subscripts denote differentiation, ρ is the density of the gas, u is the velocity, ρu is the momentum, e is the energy per unit volume, and p is the pressure. We have

$$e = \rho\epsilon + \tfrac{1}{2}\rho u^2, \tag{2d}$$

where ϵ is the internal energy per unit mass; furthermore, we assume the gas is polytropic, and thus

$$\epsilon = (1/(\gamma - 1))(p/\rho), \tag{2e}$$

where γ is a constant, $\gamma > 1$. Equations (2) are not strictly nonlinear in the sense of [5, 8], and our data will not be close to constant. Thus belief in the convergence of the method becomes at least temporarily the result of computational experience rather than the consequence of a proof.

It will be assumed in this section that a method for solving a Riemann problem and sampling the solution is available; i.e., it is assumed that given a right state S_r with $\rho = \rho_r$, $u = u_r$, $p = p_r$, $e = e_r$ in $x \geqslant 0$, and a left state S_l with $\rho = \rho_l$, $u = u_l$, $p = p_l$, $e = e_l$ in $x < 0$, the solution of Eqs. (2) can be found. This solution will consist of three states: S_r, S_l, a middle state S_* with $u = u_*$, $p = p_*$, separated by waves which may be either shocks or rarefaction waves. A slip line $dx/dt = u_*$ separates the gas initially at $x < 0$ from the gas initially at $x \geqslant 0$. u and p are continuous across the slip line, while ρ in general is not (the discontinuity of ρ would be excluded in a strictly nonlinear system). The slip line divides S_* into two parts with possibly differing values of ρ_* and e_*, but equal constant values of u_* and p_* (see Fig. 1).

Given the solution of the Riemann problem evaluated at the sample point $P = (\theta h, k/2)$, Glimm's construction can be carried out. The choice of θ of course determines the behavior of the solution. Thus, if θ is close to $-\frac{1}{2}$, the values in S_l propagate to $((i + \frac{1}{2})\,h, (n + \frac{1}{2})\,k)$, while if θ is close to $\frac{1}{2}$ the values in S_r propagate to their left.

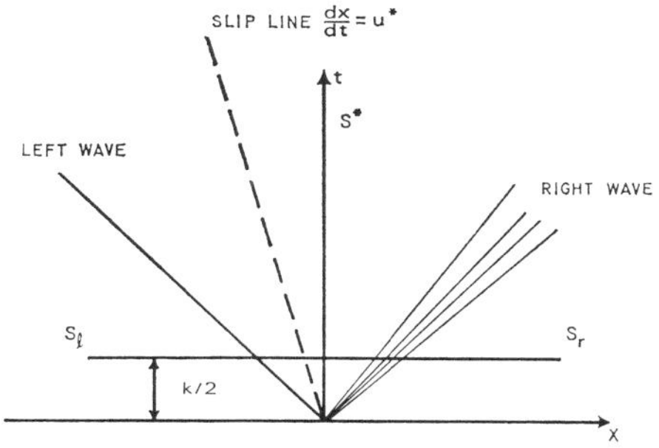

FIG. 1. Solution of Riemann problem.

If $\theta_1, \theta_2, \ldots, \theta_n, \ldots$ are the successive values of θ used in the calculation, it is important that they tend as fast as possible to approximate equipartition on $[-\frac{1}{2}, \frac{1}{2}]$. For more precise requirements, see [5]. It is natural to think of picking a new value of θ for each i and each n. The practical effect of such a choice with finite h is disastrous, except for data very close to a constant. In particular, with such a choice, there is a finite probability that a given state S will propagate to both left and right and create a spurious constant state. The first improvement in the method is a choice of a new θ only once per time level rather than once for each point and each time level. In Table I we give an example of the improvement in

TABLE I

Effect of the Sampling Procedure on the Quality of the Solution[a]

x	Run A	Run B
0.40	−1.000	−1.000
0.46	−1.000	−0.839
0.53	−0.963	−0.668
0.60	−0.963	−0.594
0.66	−0.963	−0.494
0.73	−0.963	−0.402
0.80	0.	−0.382
0.86	0.	0.

[a] Velocity field in a rarefaction wave, $t = 0.40$.

the quality of the solution which results from such a choice: Consider a gas with $\rho = 1$, $p = 1$, $u = 0$, in $0 \leqslant x \leqslant 1$. At time $t = 0$ the left boundary is set into motion with velocity $V = -1$ (we shall describe below how boundary conditions are imposed). The two columns of numbers were obtained by processes which differ only in the choice of θ. In Run A a new θ was chosen for each space interval and each time level, while in Run B a new θ was chosen for each time level only.

The variance of the solution can be further reduced by the following procedure, whose goal is to make the sequence of samples θ_i reach approximate equidistribution over $[-\frac{1}{2}, \frac{1}{2}]$ at a faster rate. Let $m_1, m_2, m_1 < m_2$, be two mutually prime (or even better, prime) integers. Consider the sequence of integers

$$n_0 \text{ given}, \quad n_0 < m_2,$$
$$n_{i+1} = (m_1 + n_i)(\text{mod } m_2).$$

(This mimics on the integers a procedure by which pseudorandom numbers are generated on the computer.) Consider the sequence of samples $\theta_1, \theta_2, \ldots$, of θ, and introduce the modified sequence

$$\theta_i' = ((n_i + \theta_i + \tfrac{1}{2})/m_2) - \tfrac{1}{2}, \qquad i = 1, 2, \ldots.$$

The θ_i' will be used instead of θ_i. To see the advantages of the modified procedure, consider a shock moving at a constant speed U between two constant states, $Uk < h$. The position X of the shock after n half-steps is given by

$$X = h \sum_{i=1}^{n} \eta_i,$$

where the η_i are random variables,

$$\begin{aligned} \eta_i &= \tfrac{1}{2} && \text{if} \quad h\theta_i < Uk/2, \\ &= -\tfrac{1}{2} && \text{if} \quad h\theta_i \geqslant Uk/2. \end{aligned}$$

The standard deviation of X is

$$h(n)^{1/2}\{(\tfrac{1}{2} + U(k/h))(\tfrac{1}{2} - U(k/h)\}^{1/2}.$$

The standard deviation is a measure of the statistical error. Its maximum value is reached when $Uk = 0$, when its value is $h(n)^{1/2}/2$. If the θ_i' are used instead of the θ_i, the maximum standard deviation in X becomes $\frac{1}{2}h(n/m_2)^{1/2}$; furthermore, if the θ_i are used, we have

$$-nh/2 \leqslant X \leqslant nh/2;$$

the extreme values $\pm nh/2$ can be reached (although with a very low probability); on the other hand, if the θ_i' are used, we have

$$m_3/m_2 \leqslant X \leqslant (m_3 + 1)/m_2$$

whenever n is a multiple of m_2, and where m_3 is the integer part of $Unkm_2/2$. When h is finite, m_2 cannot be made too large; otherwise, it introduces a systematical error into the calculation. We usually picked $m_2 = 7$, $m_1 = 3$. The convergence when $h \to 0$ is not affected as long as m_2 remains bounded.

Boundary conditions can be satisfied through the use of symmetry considerations. Consider a boundary point at $x = b$, with the fluid to the left. The boundary conditions are imposed on the grid point closest to b, say i_0h. A fake right state S_r at $(i_0 + \frac{1}{2})\,h$ is created by setting

$$\rho_{i_0+1/2} = \rho_{i_0-1/2}, \tag{3a}$$

$$u_{i_0+1/2} = -u_{i_0-1/2}, \tag{3b}$$

$$p_{i_0+1/2} = p_{i_0-1/2}. \tag{3c}$$

One then samples that part of the resulting Riemann problem which lies to the left of the slip line. If the variance reduction technique described above is used, the θ used at the boundary must be independent of the θ used in the rest of the calculation, or else there may be a finite probability that waves reflected from the boundary may never actually penetrate the interior of the domain. If an additional point $(i_0 + 1)h$ must be added to the grid, the values of ρ, u, p at this point are set equal to the values of the solution at the Riemann problem just described, sampled just to the left of the slip line.

If a passive quantity ψ is transported by the fluid. i.e., if the equation

$$\psi_t + (u\psi)_x = 0 \tag{4}$$

is added to Eqs. (2), it can be readily seen that

$$\begin{aligned} \psi_{i+1/2}^{n+1/2} &= \psi_i{}^n \qquad \text{if} \quad P = (h, k/2) \text{ is to the left of the slip line,} \\ &= \psi_{i+1}^n \qquad \text{if} \quad P \text{ is to the right of the slip line.} \end{aligned}$$

As a result, the region in which $\psi \neq 0$ remains sharply defined if it is sharply defined initially. Shocks are kept perfectly sharp. If ψ is a step function at $t = 0$, it remains a step function for all t. We express this by the statement that the method has no numerical viscosity.

The accuracy of the method cannot be assessed by usual means. There is no $n > 0$ such that the method provides an exact solution whenever that solution is a polynomial of degree n. For systems of equations with constant coefficients, the method yields a solution which exactly equals the correct solution except for a rigid translation of the coordinate system, by an amount which, with appropriate m_2, is $O(h)$. The method has then first-order accuracy but infinite resolution. In nonlinear systems, and in the presence of boundaries, the situation is more complex, and will be analyzed elsewhere.

The method is found experimentally to be unconditionally stable; in particular, it is stable when the Courant condition is violated. This fact is not paradoxical: since θ is allowed to range only over the interval $[-\frac{1}{2}, \frac{1}{2}]$, all information from outside the domain of dependence of a point is disregarded. Thus if the Courant condition is violated, the problem effectively being solved is modified.

To complete the description of the algorithm, a method for solving Riemann problems must be described. This is the object of the next section.

Solution of a Riemann Problem

The solution of a Riemann problem, needed for the application of the method just described, requires a rather lengthy explanation (and a rather lengthy program), but, since only one of the possible waves has to be fully computed, it is not uneconomical in terms of computer time.

As already described, we have at $t = 0$ two states, a left state S_l with $\rho = \rho_l$, $u = u_l$, $p = p_l$, and a right state S_r with $\rho = \rho_r$, $u = u_r$, $p = p_r$. We wish to find the solution $\bar{\rho}$, $\bar{u}$, $\bar{p}$ at the sample point $P = (\theta h, k/2)$, $-\frac{1}{2} \leqslant \theta \leqslant \frac{1}{2}$. The solution consists of S_r on the right, S_l on the left, and a state S_* in the center. S_* is separated from S_r by a right wave which is either a shock or a rarefaction, and from S_l by a left wave which is also either a shock or a rarefaction (see [4, 5, 8]).

The first step is to evaluate the pressure p_* and the velocity u_* in S_*. This is done by a method due in the main to Godunov ([6]; see also [12]). Define the quantity

$$M_r = (p_r - p_*)/(u_r - u_*). \tag{5}$$

One can show that if the right wave is a shock,

$$M_r = -\rho_r(u_r - U_r) = -\rho_*(u_* - U_r), \tag{6}$$

when ρ_* is the density in the portion of S_* adjoining the right shock and U_r is the velocity of the right shock (see [11]). In any case, one has

$$M_r = (p_r\rho_r)^{1/2}\, \phi(p_*/p_r). \tag{7}$$

where

$$\begin{aligned} \phi(w) &= \left(\frac{\gamma+1}{2}\, w + \frac{\gamma-1}{2}\right)^{1/2} && \text{for}\quad w \geqslant 1,\\ &= \frac{\gamma-1}{2(\gamma)^{1/2}}\, \frac{1-w}{1-w^{(\gamma-1)/2\gamma}} && \text{for}\quad w \leqslant 1, \end{aligned}$$

$(\phi(1) = (\gamma)^{1/2})$. Similarly, M_l is defined by

$$M_l = (p_l - p_*)/(u_l - u_*); \tag{8}$$

if the left wave is a shock,

$$M_l = \rho_l(u_l - U_l) = \rho_*(u_* - U_l), \tag{9}$$

where ρ_* is the density to the right of the left shock and U_l is the velocity of the left shock; in any case,

$$M_l = (p_l\rho_l)^{1/2}\, \phi(p_*/p_l), \tag{10}$$

where $\phi(w)$ is defined as in Eqs. (7). From (5) and (8), we have

$$p_* = (u_l - u_r + p_r/M_r + p_l/M_l)/((1/M_r) + (1/M_l)). \tag{11}$$

These considerations lead to the following iteration procedure. Pick a starting value $p_*{}^0$ (or values $M_r{}^0$, $M_l{}^0$), and then compute p_*^{q+1}, M_r^{q+1}, M_l^{q+1}, $q \geqslant 0$, using

$$\tilde{p}^q = (u_l - u_r + p_r/M_r{}^q + p_l/M_l{}^q)/((1/M_r{}^q) + (1/M_l{}^q)), \tag{12a}$$

$$p_*^{q+1} = \max(\epsilon_1, \tilde{p}^q), \tag{12b}$$

$$M_r^{q+1} = (p_r \rho_r)^{1/2}\, \phi(p_*^{q+1}/p_r), \tag{12c}$$

$$M_l^{q+1} = (p_l \rho_l)^{1/2}\, \phi(p_*^{q+1}/p_l). \tag{12d}$$

Equation (12b) is needed because there is no guarantee that in the course of iteration $\tilde{p}$ remains $\geqslant 0$. We usually set $\epsilon_1 = 10^{-6}$. The iteration is stopped when

$$\max(|\, M_r^{q+1} - M_r{}^q \,|, |\, M_l^{q+1} - M_l{}^q \,|) \leqslant \epsilon_2,$$

(we usually picked $\epsilon_2 = 10^{-6}$); one then sets $M_r = M_r^{q+1}$, $M_l = M_l^{q+1}$, and $p_* = p_{q+1}^*$.

To start this procedure one needs initial values of either M_r and M_l or p_*. The starting procedure suggested by Godunov appears to be ineffective, and better results were obtained by setting

$$p_*{}^0 = (p_r + p_l)/2.$$

We also ensured that the iteration was carried out at least twice, to avoid spurious convergence when $p_r = p_l$.

As noted by Godunov, the iteration may fail to converge in the presence of a strong rarefaction. This problem can be overcome by the following variant of Godunov's procedure. If the iteration has not converged after L iterations (we usually set $L = 20$), Eq. (12b) is replaced by

$$p_*^{q+1} = \alpha \max(\epsilon_1, \tilde{p}^q) + (1 - \alpha)\, p_*{}^q, \tag{12b$'$}$$

with $\alpha = \alpha_1 = \frac{1}{2}$. If a further L iteration occurs without convergence, we reset $\alpha_2 = \alpha_1/2$. More generally, the program was written in such a way that if the iteration fails to converge after lL iterations (l integer), α is reset to

$$\alpha = \alpha_l = \alpha_{l-1}/2.$$

In practice, the cases $l > 2$ were never encountered. The number of iterations required oscillated between 2 and 10, except at a very few points.

Once p_*, M_r, M_l are known, we have

$$u_* = (p_l - p_r + M_r u_r + M_l u_l)/(M_r + M_l)$$

from the definitions of M_r and M_l.

The fluid initially at $x \leqslant 0$ is separated from the fluid initially at $x > 0$ by a slip line whose inverse slope is $(dx/dt) = u_*$. There are now four cases to be considered:

(A) The sample point $P = (\theta h, k/2)$ lies to the left of the slip line, i.e., $\theta h \geqslant u_* k/2$, and the right wave is a shock, i.e., $p_* > p_r$;

(B) the sample point P lies to the right of the slip line and the right wave is a rarefaction, i.e., $\theta h \geqslant u_* k/2$ and $p_* \leqslant p_r$;

(C) P lies to the left of the slip line and the left wave is a shock, i.e., $\theta h < u_* k/2$ and $p_* > p_l$; and

(D) P lies to the left of the slip line and the left wave is a rarefaction, i.e., $\theta h < u_* k/2$ and $p_* \leqslant p_l$.

Case A. The velocity U_r of the right shock can be found by using Eq. (6). If the sample point P lies to the right of the shock line $dx/dt = U_r$ we have $\bar{\rho} = \rho_r$, $\bar{u} = u_r$, $\bar{p} = p_r$. If P lies to the left of the shock, $\bar{u} = u_*$, $\bar{p} = p_*$; $\bar{\rho} = \rho_*$ can be found from the second of Eqs. (6).

Case B. Let $c = (\gamma p/\rho)^{1/2}$ be the sound speed. The rarefaction is bounded on the right by the line $dx/dt = u_r + c_r$, $c_r = (\gamma p_r/\rho_r)^{1/2}$, and on the left by $dx/dt = u_* + c_*$, where $c_* = (\gamma p_*/\rho_*)^{1/2}$ can be found by using the constancy of the Riemann invariant:

$$\Gamma_r = 2c_*(\gamma - 1)^{-1} - u_* = 2c_r(\gamma - 1)^{-1} - u_r .$$

If P lies to the right of the rarefaction, $\bar{\rho} = \rho_r$, $\bar{u} = u_r$, $\bar{p} = p_r$; if P lies to the left of the rarefaction, $\bar{\rho} = \rho_*$, $\bar{u} = u_*$, $\bar{p} = p_*$. If P lies inside the rarefaction, we equate the slope of the characteristic $dx/dt = u + c$ to the slope of the line through the origin and P, obtaining

$$\bar{u} + \bar{c} = 2\theta h/k;$$

the constancy of Γ_r and the isentropic law $p\rho^{-\gamma} =$ constant, together with the definition $c = (\gamma p/\rho)^{1/2}$, yield $\bar{\rho}$, $\bar{u}$, and $\bar{p}$.

Cases C and D are mirror images of cases A and B, with M_r, Γ_r replaced by M_l and $\Gamma_l = 2c(\gamma - 1)^{-1} + u$, and will not be described in full.

Multidimensional Problems

We now generalize our version of Glimm's method to problems in two space dimensions. Problems in spaces of greater dimension can presumably be handled in a similar manner. The equations to be solved can be written in the conservation form (see, e.g., [12]):

$$\rho_t + (\rho u)_x + (\rho v)_y = 0, \tag{13a}$$

$$(\rho u)_t + (\rho u^2 + p)_x + (\rho uv)_y = 0, \tag{13b}$$

$$(\rho v)_t + (\rho uv)_x + (\rho v^2 + p)_y = 0, \tag{13c}$$

$$e_t + ((e + p)u)_x + ((e + p)v)_y = 0, \tag{13d}$$

where $\mathbf{u} = (u, v)$ is the velocity vector, x and y are the spatial coordinates, and e is the energy per unit volume, with

$$e = \rho\epsilon + \tfrac{1}{2}\rho(u^2 + v^2), \tag{13e}$$

where ϵ is the internal energy per unit mass, with the relation

$$\epsilon = (1/(\gamma - 1))(p/\rho) \tag{13f}$$

holding for polytropic gas.

The basic procedure is the use of the Glimm algorithm as a building block in a fractional step method. At each time step four quarter steps of duration $k/2$ are performed; each quarter step is a sweep in either the x or y direction. The equations to be solved in the x sweeps are

$$\rho_t + (\rho u)_x = 0, \tag{14a}$$

$$(\rho u)_t + (\rho u^2 + p)_x = 0, \tag{14b}$$

$$(\rho v)_t + (\rho uv)_x = 0, \tag{14c}$$

$$e_t + ((\rho + p)\, u)_x = 0. \tag{14d}$$

Equation (14c) can be rewritten in the (nonconservation) form

$$v_t + (uv)_x = 0, \tag{14c$'$}$$

from which it can be seen that in the x sweep v is transported as a passive scalar. Equation (14c) guarantees the conservation in the mean of v; thus Eq. (13e) can be replaced by

$$e = \rho\epsilon + \tfrac{1}{2}\rho u^2 + \text{constant}; \tag{14e}$$

the constant plays no role. Equation (13f) remains valid. Similar equations hold in the y sweeps.

There is no contradiction between the use of a splitting technique and the basic Glimm procedure. At each partial step, the solution vector is approximated by a piecewise constant vector. In the x sweeps the resulting waves in the x direction are found, and in the y sweeps the waves in the y direction are found. The task at hand is to combine the fractional steps in such a way that in the mean the interaction of the x waves and the y waves is properly accounted for.

This can be readily accomplished as follows: At the beginning of the time step ρ, p, and u are known at the points (ih, jh). After an x sweep, the solution at $((i + \frac{1}{2}) h, jh)$ is found (see Fig. 2). The solution is approximated by functions

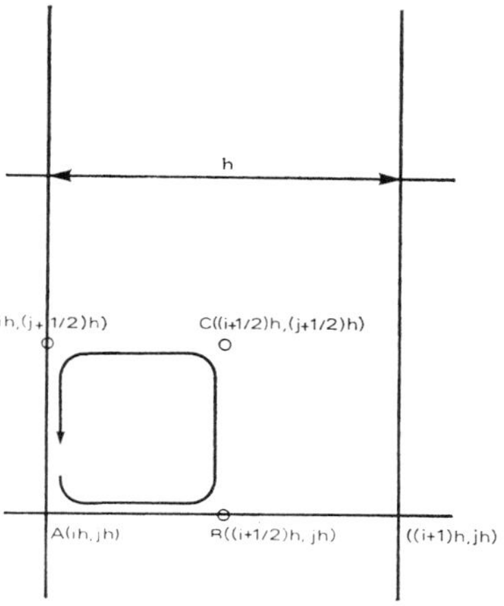

FIG. 2. Grid configuration.

constant on squares centered at these new points, and y waves are sampled, yielding a solution at $((i + \frac{1}{2}) h, (j + \frac{1}{2}) h)$. An x sweep then leads to $(ih, (j + \frac{1}{2}) h)$, and a y sweep to (ih, jh). In Fig. 2 the direction of computation is $ABCDA$. One pseudorandom number is used per quarter step.

Boundary conditions present a challenge to the application of the present method. The basic technique used is reflection. If the boundary is parallel to the mesh, no problems arise; the reflection techniques of the one-dimensional algorithm can be adapted without further ado. If the boundary lies obliquely on the grid, some suitable interpolation procedure must be found. We now describe an interpolation procedure which we found useful when the angle α between the boundary and one of the coordinate axes was small (see Fig. 3). The numerical boundary is made up of points (ih, jh) with i, j integers. Thus, boundary conditions are needed only in the third and fourth quarter steps. Let the boundary be nearly parallel to the x axis. In the third sweep, which is in the y direction, values of $\mathbf{u}$, ρ, and p are needed at points such as R, lying on a vertical line just below the boundary. Values at Q are available, and thus by reflection, values at Q' can be obtained; Q' is symmetrical to Q with respect to the boundary. At Q', ρ and p

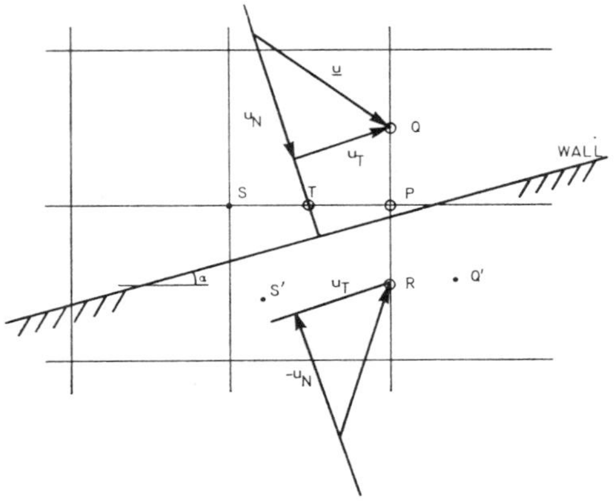

FIG. 3. Boundary conditions.

equal ρ and p at Q, the component of the velocity tangential to the boundary is the same as at Q, and the normal component has changed its sign. Similarly, values of ρ, p, and $\mathbf{u}$ at S' can be obtained by reflecting the values at S. Values at the point R can be obtained by interpolation; we can write approximately

$$\begin{aligned} \rho_R &= (1 - \tan\alpha)\,\rho_{Q'} + (\tan\alpha)\,\rho_{S'}\,, \\ p_R &= (1 - \tan\alpha)\,p_{Q'} + (\tan\alpha)\,p_{S'}\,, \\ \mathbf{u}_R &= (1 - \tan\alpha)\,\mathbf{u}_{Q'} + (\tan\alpha)\,\mathbf{u}_{S'}\,. \end{aligned} \tag{15}$$

In the last sweep, boundary conditions on the right are needed. The boundary is nearly horizontal, and thus, if (ih, j_1h), $((i+1)\,h, j_2h)$ are two neighboring boundary points, one often has $j_2 = j_1$, and thus the boundary conditions on the right do not have to be specified. If $j_2 = j_1 + 1$, then one sets ρ, p, $\mathbf{u}$ on the right equal to the values derived from (15), where $S = (ih, j_1h)$, and $Q = ((i+1)\,h, j_2)$, $j_2 = j_1 + 1$.

The method is now completely specified. We leave a discussion of the circumstances under which it is useful to the concluding section.

EXAMPLES

It is easy to identify problems in which the present method provides answers of spectacular quality; for example, in one-dimensional flow, shocks remain perfectly sharp; passive quantities are transported without diffusion in any number of dimensions, and strong shocks present no difficulties. To give a good feeling for the solutions obtained with our method, we picked a problem which is not tailored to the shape of our method; this problem involves weak shocks, oblique boundaries, discontinuities which form an angle of nearly 45$_\phi$ with the grid, and a steady state which is very sensitive to small perturbations, of just the kind the algorithm generates in abundance. It is hoped that the successful completion of the calculation will serve as a trustworthy testimonial to the power of the method.

The problem in question involves flow in a constricted channel. It was used as a test problem by Burstein [1] for two-dimensional Lax–Wendroff methods, and by Harten [7] for his artificial compression method, whose purpose is to sharpen shocks and slip lines. A supersonic flow enters a channel with a wedge angle α (see Figs. 4 and 5). In all our calculations, $\alpha = \tan^{-1}(1/5)$. When the Mach number

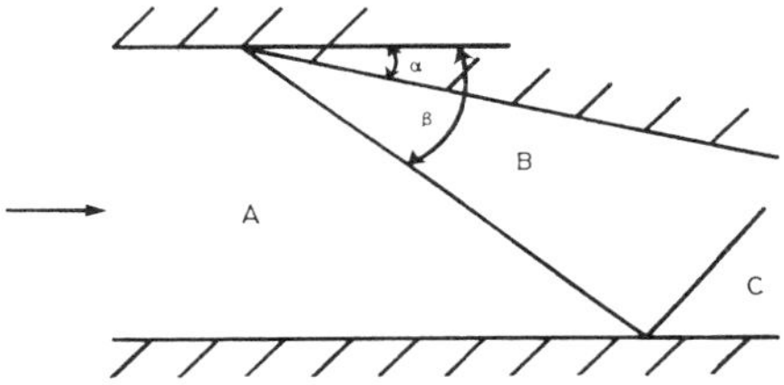

FIG. 4. Regular reflection pattern.

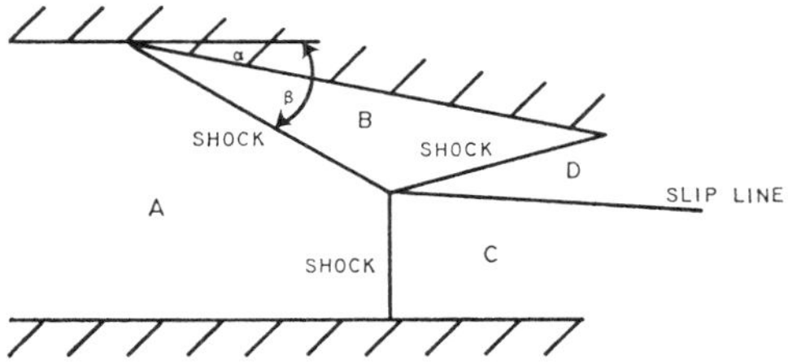

FIG. 5. Mach reflection pattern.

$M = 2$ and $\gamma = 1.4$, a regular reflection occurs at the lower boundary (Fig. 4). When $M = 1.6$ and $\gamma = 1.2$, a Mach reflection occurs (Fig. 5). In all our calculations, the flow was started impulsively, i.e., at $t = 0$ the conditions everywhere equaled the conditions on the left.

In Fig. 6 we present the density field evaluated at time $t = 6.31$, obtained with $h = 1/17$, $k = 0.0147$, $k/h = 0.25$ (the Courant condition is barely satisfied), with 17 nodes in the x direction and 12 nodes in the y direction. The running time was about 12 minutes on a CDC 6400 computer. The left boundary is maintained at the constant state $\rho = 1$, $p = 1$, $v = 0$, $u = 2(\gamma p/\rho)^{1/2} = 2.37$ (Mach number $M = 2$); $\gamma = 1.4$. At the right, the fluid is allowed to flow out freely. A shock is

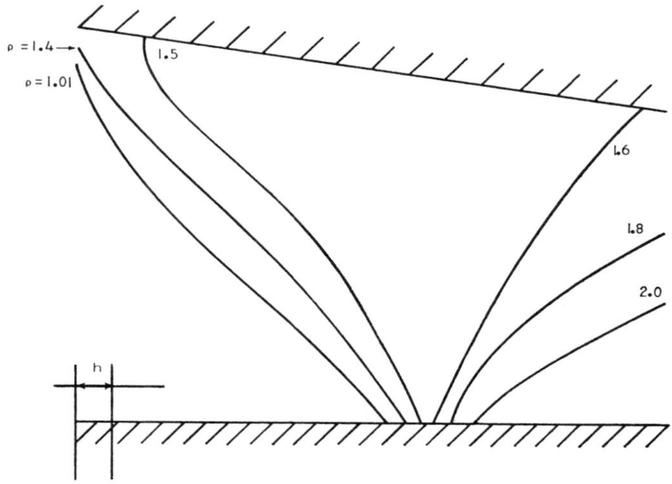

FIG. 6. Density field, Mach number $M = 2$, $\gamma = 1.4$.

induced by the wedge, and is reflected at the lower boundary. The shock angle β (Fig. 4) should be 40.8 degrees; the computed shock angle is indistinguishable from this value; it is of course not computed accurately on such a crude grid. The exact value of ρ in region B is $\rho = 1.52$, and in region C, $\rho = 2.27$ (see [4]). In Table II we give the values of ρ along the lower boundary. The position of the shocks is marked by an arrow.

In Fig. 7 we display the density field obtained with $M = 1.6$, $\gamma = 1.2$. A Mach reflection occurs. In region B the exact value of ρ is 1.48. In region A we have of couse $\rho = 1$. We used $h = 1/14$, $k/h = 0.25$, a grid of 17×12, and ran 184 time steps. Since the upstream flow is subsonic, small perturbations affect the calculation and on the right the steady state has not been reached. With good will, the slip line can be seen.

These results are at least equal in quality to those obtained previously, and much superior to those one can expect from any straightforward first-order method (see [6]). It can be seen that the steady state is never fully achieved. The correct shock transition occurs; it is built into the method, and besides, the method conserves energy, momentum, and mass in the mean. One feature of the results obtained is that they contain fluctuation of a small scale and amplitude. This can be eliminated through the use of a small artificial viscosity (the present method has none), but such a viscosity would destroy one of the most important advantages of the method. Besides, a quick glance at a real flow field will show that small fluctuations hardly detract from the physical meaning of the computed results.

TABLE II

Density at Lower Wall[a]

$t = 5.30$ ($n = 360$)	$t = 6.32$ ($n = 430$)	
1.00	1.00	exact $\rho = 1$
1.00	1.00	
1.00	1.00	
1.01	1.00	
1.02	1.00	
1.08	1.51	
1.26	1.85	exact position of ← triple point
1.58	1.92	exact $\rho = 2.27$
2.04	2.38	
2.41	2.60	
2.22	2.49	
2.24	2.38	

[a] Mach No. $M = 2$, $\gamma = 1.6$, regular reflection, 17×12 grid.

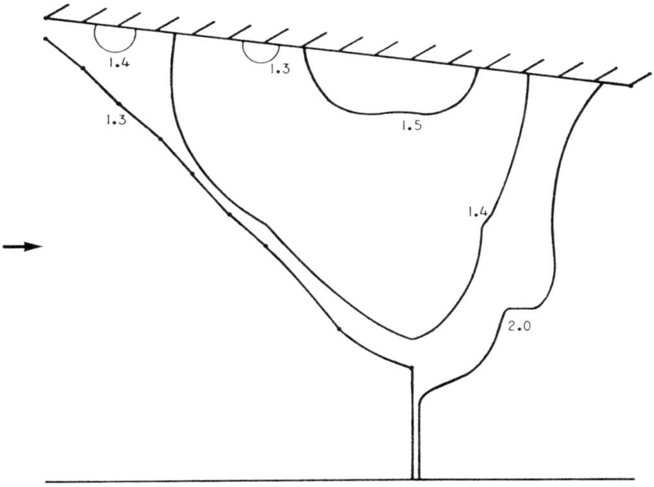

FIG. 7. Density field, $M = 1.6$, $\gamma = 1.2$.

Conclusions

We have developed a random choice method for solving the equation of gas dynamics. This method is obviously not competitive with more classical methods when it is applied to problems whose solutions are smooth, since in such problems greater accuracy can be achieved with much smaller effort. The method is destined for use in problems which involve complex patterns of discontinuity; in such problems, the greater effort required per mesh point is balanced by the economy in representation, which requires fewer points per problem.

The interesting features of this explicit method are its unconditional stability, and its neglect of all characteristic velocities larger than h/k. Those are features one desires to obtain, not always successfully, in implicit methods. The situations in which these features are particularly desirable are those in which the equations have multiple significant scales in either space or time; this happens for example in combustion problems, in some problems involving two phase flow, and in problems where both boundary layers and shocks play a significant role (see, e.g., [3]).

It may be interesting to compare the present method with the random vortex method [2] where a random choice is also an essential feature. In both methods the random choice feature is used to control the numerical dissipation. However, in the random vortex method the random choice is used to represent the real viscosity as well as control the numerical viscosity. This fact makes it imperative that the random numbers picked at the various spatial locations at a fixed time be independent. In the present method this constraint does not apply and, in fact, it is essential that it be flouted.

Finally, it should be stressed that a certain randomness is a property of many real flows, and thus a method which exhibits randomness is not necessarily less desirable than a method which yields fully predictable answers.

Note. The programs used to obtain the results above are available from the author.

Acknowledgments

I would like to thank Mr. Phillip Colella and Mr. William Noh for a number of helpful discussions and comments, and Mr. Phillip Colella also for his assistance in programming and in checking calculations.

References

1. S. Z. Burstein, *AIAA J.* **2** (1964), 211.
2. A. J. Chorin, *J. Fluid Mech.* **57** (1973), 785.

3. A. J. Chorin, A numerical method for studying the dynamical effects of exothermic reactions, to appear.
4. R. Courant and K. O. Friedrichs, "Supersonic Flow and Shock Waves," Interscience, New York, 1948.
5. J. Glimm, *Comm. Pure Appl. Math.* **18** (1965), 697.
6. S. K. Godunov, *Mat. Sb.* **47** (1959), 271.
7. A. Harten, The method of artificial compression, AEC R & D report C00-3077-50, New York University, 1974.
8. P. D. Lax, *Comm. Pure Appl. Math.* **10** (1957), 537.
9. P. D. Lax, *SIAM Rev.* **11** (1969), 7.
10. C. Moler and J. Smoller, *Arch. Rational Mech. Anal.* **37** (1970), 309.
11. A. K. Oppenheim, "Introduction to Gas Dynamics of Explosions," Lectures in Mechanical Sciences, Springer, New York, 1970.
12. R. D. Richtmyer and K. W. Morton, "Finite Difference Methods for Initial Value Problems," Interscience, New York, 1967.

Random Choice Methods with Applications to Reacting Gas Flow

ALEXANDRE JOEL CHORIN*

Department of Mathematics and Lawrence Berkeley Laboratory, University of California, Berkeley, California 94720

Received October 28, 1976; revised February 1, 1977

The random choice method is analyzed, appropriate boundary conditions are described, and applications to time-dependent reacting gas flow in one dimension are carried out. These applications illustrate the advantages of the method when one is solving problems where the diffusion constant vanishes or is very small, and where artificial or numerically induced diffusion cannot be tolerated.

INTRODUCTION

The random choice method for solving hyperbolic systems was introduced as a numerical tool in [2]. It is based on a constructive existence proof due to Glimm [5]. In this method, the solution of the equations is constructed as a superposition of locally exact elementary similarity solutions; the superposition is carried out through a sampling procedure. The computing effort per mesh point is relatively large, but the global efficiency is high when the solutions sought contain components of widely differing time scales. This efficiency is due to the fact that the appropriate interactions can be properly taken into account when the elementary similarity solutions are computed. The aim of the present paper is to provide a further analysis of the method, and to illustrate its usefulness in the analysis of reacting gas flow. Examples are given of time-dependent detonation and deflagration waves, with infinite and finite reaction rates.

We begin by describing the method briefly. Consider the hyperbolic system of equations

$$\mathbf{v}_t = (\mathbf{f}(\mathbf{v}))_x, \qquad \mathbf{v}(x, 0) \quad \text{given}, \tag{1}$$

when $\mathbf{v}$ is the solution vector and subscripts denote differentiation. The time t is divided into intervals of length k. Let h be a spatial increment. The solution is to be evaluated at the points (ih, nk) and $((i+\frac{1}{2})h, (n+\frac{1}{2})k)$, $i = 0, \pm 1, \pm 2,\ldots, n = 1, 2,\ldots$. Let $\mathbf{u}_i^n$ approximate $\mathbf{v}(ih, nk)$, and $\mathbf{u}_{i+1/2}^{n+1/2}$ approximate $\mathbf{v}((i+\frac{1}{2})h, (n+\frac{1}{2})k)$. The

* Partially supported by the Office of Naval Research under Contract No. N00014-76-C-0316, and by the U.S. ERDA.

Reprinted with permission from *Journal of Computational Physics*, Vol. 25, No. 3, 1977.

algorithm is defined if $\mathbf{u}_{i+1/2}^{n+1/2}$ can be found when $\mathbf{u}_i^n$, $\mathbf{u}_{i+1}^n$ are known. Consider the following Riemann problem:

$$\mathbf{v}_t = (\mathbf{f}(\mathbf{v}))_x, \qquad t > 0, -\infty < x < +\infty,$$

$$\mathbf{v}(x, 0) = \mathbf{u}_{i+1}^n \qquad \text{for} \quad x > 0,$$

$$= \mathbf{u}_i^n \qquad \text{for} \quad x < 0.$$

Let $\mathbf{w}(x, t)$ denote the solution of this problem. Let θ_i be a value of a variable θ, $-\frac{1}{2} \leqslant \theta \leqslant \frac{1}{2}$. Let P_i be the point $(\theta_i h, k/2)$, and let

$$\tilde{\mathbf{w}} = \mathbf{w}(P_i) = \mathbf{w}(\theta_i h, k/2)$$

be the value of the solution $\mathbf{w}$ of the Riemann problem at P_i. We set

$$\mathbf{u}_{i+1/2}^{n+1/2} = \tilde{\mathbf{w}}.$$

In other words, at each time step, the solution is first approximated by a piecewise constant function; it is then advanced in time exactly, and new values on the mesh are obtained by sampling. The usefulness of the method depends on the possibility of solving Riemann problems efficiently.

Simple Examples and Partial Error Estimates

In order to explain the method further and analyze its limitations, in this section we consider simple examples of its use; the first one was already discussed in [7]. Consider the equation

$$v_t = v_x \tag{2}$$

in $-\infty < x < +\infty$, $t > 0$, with $v(x, 0) = g(x)$ given. One can readily see that if a single θ is picked per half time step, Glimm's method reduces to

$$u_{i+1/2}^{n+1/2} = u_{i+1}^n \qquad \text{if} \quad \theta h \geqslant -k/2,$$

$$= u_i^n \qquad \text{if} \quad \theta h < -k/2.$$

It follows that

$$u_i^n = v(ih + \eta, t),$$

where $\eta = \eta(t)$ is a random variable which depends on t alone; i.e., the computed solution equals the exact solution with a shift independent of x. The magnitude of η depends on the choices of θ. Consider the following strategies for picking θ:

(i) θ is picked at random from the uniform distribution on $[-\frac{1}{2}, \frac{1}{2}]$;

(ii) n is assumed known in advance; the interval $[-\frac{1}{2}, \frac{1}{2}]$ is divided into n subintervals of equal lengths and θ_i is picked in the middle of the ith subinterval;

(iii) (a compromise between (i) and (ii)): $[-\frac{1}{2}, \frac{1}{2}]$ is divided into m subintervals, $m \ll n$, and θ_1 is picked at random in the first subinterval, θ_2 in the second subinterval, θ_{m+1} in the first subinterval, etc.

A fourth strategy which relies on the well-equipartitioned sequences studied by Richtmyer and Ostrowski was suggested by Lax [7], but is not useful in the present context.

If strategy (i) is used, we have

$$\begin{aligned} x + \eta &= \text{displacement of the initial value} \\ &= \sum_{i=1}^{2n} \eta_i , \end{aligned}$$

where

$$\begin{aligned} \eta_i &= h/2 \quad \text{if} \quad h\theta_i < -k/2, \\ &= -h/2 \quad \text{if} \quad h\theta_i \geqslant -k/2. \end{aligned}$$

The variance of η_i is readily evaluated:

$$\operatorname{var}(\eta_i) = \frac{h^2}{4}\left(1 - \frac{k}{h}\right)\left(1 + \frac{k}{h}\right);$$

the variance of η is thus

$$\frac{nh^2}{4}\left(1 - \frac{k}{h}\right)\left(1 + \frac{k}{h}\right),$$

and the standard deviation of η, which measures its magnitude, is $(n^{\frac{1}{2}}h/2)\,\{(1 - (k/h))(1 + (k/h))\}^{\frac{1}{2}} = O(n^{\frac{1}{2}}h)$.

If the second strategy is used,

$$u_i^n = v(x + \eta, t), \qquad |\,\eta\,| \leqslant 1/2n,$$

if $n = O(h^{-1})$, $\eta = O(h)$. If the third strategy is used, and n is a multiple of m, $\eta = O(h(n/m)^{\frac{1}{2}})$, since only in every mth half step is the outcome of the sampling in doubt.

Assume v is of compact support. Following a suggestion by Lax, we define the resolution of the scheme by

$$Q^{-1} = \min_q \|\, u_i^n - v(ih + q, t)\|,$$

where $\|\ \|$ denotes the maximum norm. The scheme has resolution of order m if $Q = O(h^{-m})$. The displacement d of the scheme is defined by

$$\begin{aligned} Q^{-1} &= \|\, u_i^n - v(ih + d, t)\| \\ &= \min_q \|\, u_i^n - v(ih + q, t)\|. \end{aligned}$$

The method applied to the present problem has almost first-order accuracy, almost first-order displacement, but infinite resolution. There is no smoothing and no numerical diffusion or dispersion. For any k/h, the domain of dependence of a point is always a single point. The answers are always bounded. If the Courant condition $k/h \leqslant 1$ is violated, the equation being approximated is $v_t = (h/k)v_x$. Clearly, since these results are independent of k/h, they generalize to hyperbolic systems with constant coefficients.

Consider now the equation

$$v_t = a(x, t)v_x ,$$

in $-\infty < x < +\infty$, $t > 0$, $v(x, 0) = g(x)$ given, and $a(x, t)$ a Lipschitz continuous function of both x and t. The method is not well suited to the solution of such an equation, both because the solution of the Riemann problem requires a possibly laborious integration of a characteristic equation, and because the errors will turn out to be large compared with those incurred in other available methods. The analysis is nevertheless illuminating.

Let C_{x_0} be the characteristic

$$dx/dt = -a(x, t), \qquad x(0) = x_0 .$$

For each i, we have

$$u_{i+1/2}^{n+1/2} = u_{i+1}^{n} \quad \text{if} \quad P = (\theta h, k/2) \text{ lies to the right of } C_{(1+\frac{1}{2})h} ,$$

$$= u_i^n \quad \text{if} \quad P \text{ lies to the left of } C_{(i+\frac{1}{2})h} .$$

As before,

$$u_i^n = v(x + \eta, t), \qquad x = ih, \qquad t = nk;$$

where η is a random variable which now depends on both x and t.

If θ is picked at random from the uniform distribution on $[-\frac{1}{2}, \frac{1}{2}]$ (strategy (i)) we have as before $\eta = O(hn^{\frac{1}{2}})$. Strategy (ii) clearly yields an error $O(1)$. Strategy (iii) is more advantageous; the standard deviation of η is again bounded by $O(h(n/m)^{\frac{1}{2}})$. However, the mean of η is no longer zero. Assume $k = O(h)$. Note that $a(x, t)$ may vary by $O(mh)$ before this change affects the values of η. Thus, $\bar{\eta} =$ mean of $\eta = O(mh)$, and $\eta = O(mh) + O(h(n/m)^{\frac{1}{2}})$. If $n = O(h^{-1})$ and $m = O(n^{\frac{1}{3}})$, then $\eta = O(h^{\frac{2}{3}})$. We have less than first-order accuracy and more than first-order displacement.

We now try to assess the relative displacement of two points. Let us assume that the first sampling strategy is used, i.e., θ is picked at each step from the uniform distribution on $[-\frac{1}{2}, \frac{1}{2}]$. Consider first the quantity

$$\Delta\eta(h, k) = (\eta(x, t + k) - \eta(x + h, t + k)) - (\eta(x, t) - \eta(x + h, t)),$$

i.e., the difference between the numerically induced translations experienced by two neighboring points during one time step. If $\Delta\eta(h, k) > 0$, information is lost: one value of $v(x, 0)$ disappears. If $\Delta\eta(h, k) < 0$, a false constant state is created. $\Delta\eta(h, k)$

can take on the values 0, $\pm h$. $\Delta\eta(h, k) \neq 0$ if $P = (\theta h, k/2)$ falls to the left of the characteristic through one of the points (ih, nk), $((i+1)h, nk)$ and to the right of the other. This happens with probability $O(h)$. That is, the variance of $\Delta\eta(h, k)$ is $O(h^3)$. Therefore, the variance of $\Delta\eta(h) = \eta(x, t) - \eta(x+h, t)$ is $nO(h^3) = O(h^2)$ if $n = O(h^{-1})$, and the standard deviation of $\Delta\eta(h)$ is $O(h)$, i.e., neighboring values in the range of v do not fly far apart. The same estimate holds for the other sampling strategies.

Consider now the relative displacement $\Delta\eta$ of two values far apart. Let $\eta_1 = \eta(x, t)$, $\eta_2 = \eta(x + X, t)$, and $\Delta\eta = \eta_2 - \eta_1$, and thus

$$u_i^n = v(x + \eta_1, t) = g(x_1), \qquad ih = x, \qquad nk = t,$$

$$u_{i+i_0}^n = v(x + X + \eta_2, t) = g(x_2), \qquad i_0 h = X,$$

where $g(x) = v(x, 0)$. Let C_{x_1} be the characteristic through $(x_1, 0)$, and similarly for C_{x_2}. $x_2 - x_1$ has increased by $\pm h$ each time $P = (\theta h, k/2)$ fell between the two characteristics. Assume the first sampling strategy is used. There are two sources of error which make $\Delta\eta \neq 0$. There is the standard deviation of the sum of the random variables which equal $\pm h$ when P is between the characteristics, and are zero otherwise (this is clearly $O(hn^{\frac{1}{2}})$), and there is the uncertainty in the slope of the characteristics due to the lateral displacement of the solution; this is again $O(hn^{\frac{1}{2}})$ and induces an error $O(h^{\frac{3}{2}}n^{\frac{3}{4}}) = O((hn^{\frac{1}{2}})^{\frac{3}{2}})$; if $n = O(h^{-1})$, this is $O(h^{\frac{3}{4}})$. Thus $\Delta\eta = O(hn^{\frac{1}{2}})$, and the resolution is not of higher order than the accuracy. Similar results hold for the other sampling strategies.

We now turn to the nonlinear problem

$$\mathbf{v}_t = (f(\mathbf{v}))_x,$$

where f is a function of v but not explicitly a function of x and t. The method of analysis we have used here is not applicable, since values of $\mathbf{v}$ are not merely propagated along characteristics. Furthermore, we have here no way of taking into account properly the fact that rarefaction or loss of information incurred in the numerical process correspond to genuine properties of the differential equations. All we can provide here is a heuristic analysis. Consider the third sampling strategy. Since the slope of the characteristic depends on the values of v and not on x, and values of v at neighboring points remain attached to neighboring points, we expect the term $O(mh)$ in η to disappear, and have $\eta = O(h(n/m)^{\frac{1}{2}})$. Thus, the resolution should be at least $O(h(n/m)^{\frac{1}{2}})$. Note that if $n = O(h^{-1})$ and $m = O(n)$, the random element in the method loses its significance.

In the case of a shock separating two constant states, one can readily see that $d = O(h(n/m)^{\frac{1}{2}})$ but the resolution is infinite. One can trivially define resolution in a neighborhood. Thus, what we have is a rather awkward first-order method, which resolves shocks very sharply. We also know that it keeps fluid interfaces perfectly sharp [2]. It is useful for the analysis of problems in Cartesian coordinates in which the dynamics of the discontinuities are of paramount significance. We provide examples of such problems in later sections. Recent results (see, e.g., [8]) show that in such problems substantially higher accuracy cannot be achieved.

Boundary Conditions

The correct imposition of boundary conditions in our method requires careful thought, and was not adequately discussed in [2]. It is clear that even in the case of Eq. (2) the presence of a boundary can detract from both accuracy and resolution. The lateral displacement of the solution may make some function values disappear across the boundary and care must be taken to ensure the possibility of their retrieval. Additional storage across the boundary and careful accounting of the lateral displacement provide a remedy.

The following procedure has been introduced in [2] to reduce the lateral displacement of the solution (and thus reduce the loss of information at walls), when the third sampling strategy is used. The goal is to obtain as fast as possible solution values on both sides of whatever wave pattern emerges in the solution of the Riemann problem, and thus rapidly offset a displacement to right by a displacement to the left (or vice versa). We pick an integer $m' < m$, m and m' mutually prime, and n_0 integer, $n_0 < m$, and construct the sequence of integers

$$n_{i+1} = (n_i + m') \pmod{m}. \tag{3}$$

The subintervals of $[-\frac{1}{2}, \frac{1}{2}]$ are then sampled in the order $n_0, n_1, n_2, \ldots$ rather than in the natural succession. One can furhter modify the sampling so that of two successive values of θ, one lies in $[-\frac{1}{2}, 0]$ and one in $[0, \frac{1}{2}]$. These procedures do not increase the error far from the wall, and are quite effective, although no analytical assessment of their efficiency is available.

Suppose we are solving the equations of gas dynamics (Eqs. (4) below), and using the third sampling strategy, modified by (3) or not. Assume the velocity v is given at the boundary. One can find a state (i.e., a set of values for the gas variables) which has the given velocity and which can be connected to the state one mesh point into the fluid by a simple wave (see, e.g., [4]). This is equivalent to solving half a Riemann problem, and provides an appropriate solution field which can be sampled. The same result can be obtained by symmetry considerations. Consider a boundary point to the right on the region of flow; let the boundary conditions be imposed at a point i_0h. A fake right state at $(i_0 + \frac{1}{2})h$ is created, with

$$\rho_{i_0+1/2} = \rho_{i_0-1/2},$$

$$v_{i_0+1/2} = 2V - v_{i_0-1/2},$$

$$p_{i_0+1/2} = p_{i_0-1/2},$$

where ρ, v, p are, respectively, the gas density, velocity, and pressure, and V is the velocity of the wall. The constant state in the middle of the Riemann solution is the wall state, and it is sampled to the left of the slip line $dx/dt = V$.

This procedure contains a pitfall, not noticed in [2]; let θ be chosen in accordance with our usual sampling strategy; let θ_1, θ_2 be the values of θ at two successive time steps (θ_1 and θ_2 are not independent). θ_1', θ_2', the values used at the wall, differ from θ_1 and θ_2 since only part of the interval $[-\frac{1}{2}, \frac{1}{2}]$ is sampled (or else one does not remain to the left of the wall line $dx/dt = V$). θ_1' and θ_2' can presumably be obtained by a linear change of variables. Consider a specific part of the wave pattern at the wall. Since θ_1', θ_2' are not independent, the possibility exists that whenever θ_1' picks up the specific part we are considering, θ_2' is such that this information is lost to the wall. This possibility was not noticed in [2], and its removal by the methods whose description follows contributes to the sharpening of the results obtained in [2].

It is always consistent to pick θ_1', θ_2' by a linear change of variables from two values picked independently from the uniform distribution on $[-\frac{1}{2}, \frac{1}{2}]$. On the average no information will be lost to the wall, but the variance of the solution will be increased. Better strategies can be devised, but require thought in each special case. If the walls are at rest, $V = 0$, one can proceed as follows: impose the boundary condition on the right at time nk and a point $i_1 h$, and on the left at time $(n + \frac{1}{2})k$ at a point $(i_2 + \frac{1}{2})h$, i_1, i_2 integers. One can see that if θ_1, θ_2 are so chosen that $\theta_1 \leqslant 0$ at time nk, and $\theta_2 \geqslant 0$ at time $(n + \frac{1}{2})k$, then θ_1 and θ_2 can be used at the boundary as well as in interior without loss of resolution.

Detonations and Deflagrations in a One-Dimensional Ideal Gas

Our goal in this section is to present a quick summary of the elementary theory of one-dimensional detonation and deflagration waves (for more detail, see, e.g., [4; 10]), and then derive some relations between the hydrodynamical variables on the two sides of such waves for later use.

The equations of gas dynamics are

$$\rho_t + (\rho v)_x = 0, \tag{4a}$$

$$(\rho v)_t + (\rho v^2 + p)_x = 0, \tag{4b}$$

$$e_t + ((e + p)v)_x = 0, \tag{4c}$$

where the subscripts denote differentiation, ρ is the density of the gas, v is the velocity, ρv is the momentum, e is the energy per unit volume, and p is the pressure. We have

$$e = \rho\epsilon + \tfrac{1}{2}\rho v^2, \tag{4d}$$

where $\epsilon = \epsilon_i + q$, ϵ_i is the internal energy per unit mass,

$$\epsilon_i = \frac{1}{\gamma - 1}\frac{p}{\rho}, \tag{4e}$$

where γ is a constant, $\gamma > 1$, and q is the energy of formation which can be released through chemical reaction (see [4]). In the present section it is assumed that part of q is released instantaneously in an infinitely thin reaction zone. Let the subscript 0 refer to unburned gas (i.e., gas which has not yet undergone the chemical reaction) and let the subscript 1 refer to burned gas. The unburned gas is on the right. We have

$$\epsilon_1 = \frac{1}{\gamma_1 - 1}\frac{p_1}{\rho_1} + q_1,$$

$$\epsilon_0 = \frac{1}{\gamma_0 - 1}\frac{p_0}{\rho_0} + q_0.$$

For the sake of simplicity, here we make the unrealistic assumption $\gamma_1 = \gamma_0 = \gamma$. (The case $\gamma_1 \neq \gamma_0$ is more difficult only because of additional algebra.) When $\gamma_1 = \gamma_0 = \gamma$ the reaction can be exothermic (i.e., release energy) only if $q_0 > q_1$.

Let U be the velocity of the reaction zone. Let

$$w_1 = v_1 - U, \qquad w_0 = v_0 - U.$$

Conservation of mass and momentum is expressed by

$$\rho_1 w_1 = \rho_0 w_0 = -M, \tag{5}$$

$$\rho_0 {w_0}^2 + p_0 = \rho_1 {w_1}^2 + p_1 \tag{6}$$

(see [4]). From these relations one readily deduces

$$M^2 = -(p_0 - p_1)/(\tau_0 - \tau_1), \qquad \text{where} \quad \tau = 1/\rho.$$

Define the function H by

$$H = \epsilon_1 - \epsilon_0 + ((\tau_0 - \tau_1)/2)\,(p_1 + p_0).$$

Conservation of energy is expressed by

$$H = H(\tau_1, p_1, \tau_0, p_0) = 0.$$

Define $\Delta = q_1 - q_0$ ($\Delta \leqslant 0$ for an exothermic process), and $\mu^2 = (\gamma - 1)/(\gamma + 1)$; we find

$$\begin{aligned} 2\mu^2 H = 0 &= (1 - \mu^2)\,\tau_0 p_0 - (1 - \mu^2)\,p_1\tau_1 - 2\mu^2\Delta + ((\tau_0 - \tau_1)/2)(p_1 + p_0) \\ &= p_0(\tau_0 - \mu^2\tau_1) - p_1(\tau_1 - \mu^2\tau_0) - 2\mu^2\Delta. \end{aligned} \tag{7}$$

In the (τ_1, p_1) plane the locus of points which can be connected to (τ_0, p_0) by an infinitely thin combustion wave is a curve which reduces to a hyperbola when Δ is independent of p and τ. (See Fig. 1.) The lines through (τ_0, p_0) tangent to $H = 0$ are called the Rayleigh lines. Their points of tangency, S_1 and S_2, are called the

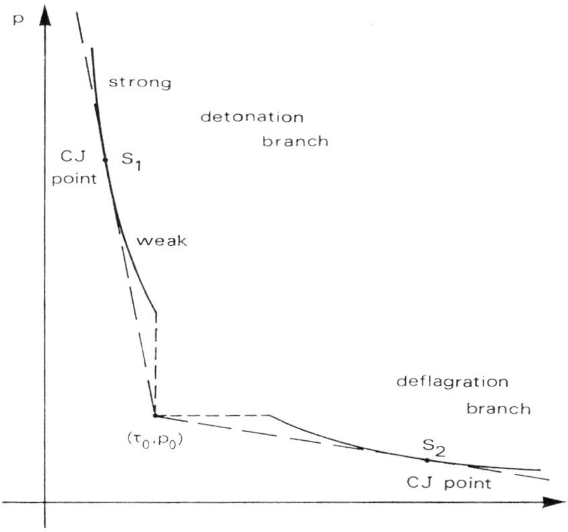

FIG. 1. The Hugoniot curve for exothermic gas flow.

Chapman–Jouguet (CJ) points. A portion of the curve is omitted because it corresponds to unphysical events in which $M^2 < 0$. The upper portion of the curve corresponds to detonations; the portion above S_1 to strong detonations and the portion below to weak detonations. The lower part of the curve corresponds to deflagrations.

The velocity and strength of a strong detonation are entirely determined by the state of the unburned gas in front of the detonation and one quantity behind the detonation, just as in the case with shocks. Let p_0, ρ_0, τ_0, ϵ_0, and v_0 be given, as well as p_1, and assume the unburned gas lies to the right of the detonation. We have from (7)

$$\tau_1 = \tau_0 \left(\frac{p_0 + \mu^2 p_1}{\mu^2 p_0 + p_1} \right) + \frac{2\mu^2 \Delta}{\mu^2 p_0 + p_1}, \tag{8}$$

and thus

$$M^2 = \frac{p_0 - p_1}{\tau_0 - \tau_1} = (p_0 - p_1) \Big/ \tau_0 \left(\frac{p_0 + \mu^2 p_1}{p_1 + \mu^2 p_0} + \frac{2\mu^2 \Delta \rho_0}{\mu^2 p_0 + p_1} - 1 \right);$$

Let $[p] = p_1 - p_0$; some algebra yields

$$M^2 = p_0 \rho_0 \left(\frac{\gamma - 1}{2} + \frac{\gamma + 1}{2} \left(\frac{p_1}{p_0} \right) \right) \Big/ (1 - (\gamma - 1)\, \rho_0 \Delta / [p]). \tag{9}$$

If $\Delta = 0$ this formula reduces to the expression for M in a shock, as given in [2] or [9]. M is real if $[p] - (\gamma - 1)\, \rho_0 \Delta \geq 0$; this can be readily seen to hold in a strong detonation.

The states on the curve $H = 0$ located between the CJ point S_1 and the line $\tau = \tau_0$ correspond to weak detonations. As described in [4], the state behind a weak detonation is entirely determined by the velocity U of the detonation and the state in front of it. In fact, a weak detonation cannot occur and what does happen is a CJ detonation followed by a rarefaction wave. Our next objective is to derive an explicit criterion for determining whether a detonation will be a strong detonation or a CJ detonation.

It is shown in [4] that at S_1, $|w_1| = c_1$ where $c_1 = (\gamma p_1/\rho_1)^{\frac{1}{2}}$ is the sound speed, i.e., a CJ detonation moves with respect to the burned gas with a velocity equal to the velocity of sound in the burned gas. We now use this fact to determine the density ρ_{CJ}, velocity v_{CJ}, and pressure p_{CJ} behind a CJ detonation.

From Eqs. (4) and (5) one finds

$$(p_1 - p_0)/(\tau_1 - \tau_0) = -\rho_1^2 w_1^2 = -\rho_0^2 w_0^2 = -M^2,$$

and thus in a CJ detonation

$$\frac{p_1 - p_0}{\tau_1 - \tau_0} = -\rho_1^2 \frac{\gamma p_1}{\rho_1} = \frac{-\gamma p_1}{\tau_1}, \qquad \tau_1 = \frac{1}{\rho_1},$$

or

$$\tau_1(p_1(1+\gamma) - p_0) = \gamma \tau_0 p_1 . \tag{10}$$

Equating τ_1 obtained from (8) to τ_1 in (10), we find

$$\tau_0 \left(\frac{\mu^2 p_1 + p_0}{p_1 + \mu^2 p_0}\right) + \frac{2\mu^2 \Delta}{(p_1 + \mu^2 p_0)} = \frac{\gamma \tau_0 p_1}{p_1(1+\gamma) - p_0}.$$

Some algebra reduces this equation to

$$p_1^2 + 2p_1 b + c = 0,$$

where

$$b = -p_0 - 2\Delta(\gamma - 1)\rho_0, \tag{11a}$$

$$c = p_0^2 + 2\mu^2 p_0 \rho_0 \Delta; \tag{11b}$$

a trivial calculation shows that $b^2 - c \geqslant 0$ if $\gamma \geqslant 1$ and $\Delta \leqslant 0$. Thus

$$p_{CJ} = p_1 = -b + (b^2 - c)^{\frac{1}{2}}, \tag{11c}$$

where the $+$ sign is mandatory since a detonation is compressive. (One can verify that if the $-$ sign were chosen, we would have $p_{CJ} < p_0$, i.e., the detonation would not be compressive. With the $+$ sign, $p_{CJ} > p_0$.) Given $p_{CJ} = p_1$, $\rho_{CJ} = \rho_1 = \tau_1^{-1}$ can be obtained from Eq. (10). Since $M = -\rho_1 w_1$, and $w_1 = -c_1$, we find

$$M = (\gamma p_1 \rho_1)^{\frac{1}{2}} = (\gamma p_{CJ} \rho_{CJ})^{\frac{1}{2}}.$$

The velocity U_{CJ} of the detonation is found from

$$\rho_0(v_0 - U_{CJ}) = -M,$$

which yields $U_{CJ} = (\rho_0 v_0 + (\gamma p_{CJ} \rho_{CJ}))/\rho_{CJ}$, and then

$$v_{CJ} = U_{CJ} - c_{CJ} . \tag{12}$$

v_{CJ} depends only on the state of the unburned gas.

Suppose v_1, the velocity of the burned gas, is given. If $v_1 \leqslant v_{CJ}$ a CJ detonation appears, followed by a rarefaction wave. If $v_1 = v_{CJ}$ a CJ detonation appears alone, and if $v_1 > v_{CJ}$ a strong detonation takes place.

If the unburned gas lies to the left of the burned gas analogous relations are found; the only difference lies in the signs of v, in particular,

$$M = +\rho_1(v_1 - U) = +\rho_0(v_0 - U).$$

The velocity of a possible deflagration cannot be determined within the context of a theory which assumes the gas to be nonconducting; this point is discussed further below. It turns out that for a nonconducting gas the only possible deflagration is a constant pressure deflagration, $p_1 = p_0$, which moves with zero velocity with respect to the gas; i.e., it is indistinguishable from a slip line.

APPLICATION OF THE METHOD TO REACTING GAS FLOW

One interesting feature of our method is its applicability to the analysis of gas flow in which exothermic chemical reactions are taking place and producing substantial dynamical effects. A Riemann problem is solved at each time step and at each point in time; this solution is then sampled. The advantage of this procedure is that the interaction of the flow and the chemical reaction can be taken into account when the Riemann problem is solved, even when the time scales of the chemistry and the fluid flow are very different. As a result, the basic conservation laws are satisfied at the end of each time step. It can be readily seen that if the chemical reactions and the gas flow were to be taken into account in separate fractional steps, the basic conservation laws may be violated at the end of each hydrodynamical step, thus either inducing unwanted oscillations and waves, or requiring time steps small enough for all changes to be very gradual—usually a costly remedy. It is interesting to note that the Riemann solutions with energy deposition in the flow field are equivalent to the exothermic centers introduced by Oppenheim [3] and serve the same purpose of accounting for the dynamical effects of the exothermic reactions. These discrete exothermic centers correspond to a physical reality whose origin can be ascribed to the fluctuations in the levels of chemical species [1].

We consider here the simplest possible description of a reacting gas (see, e.g., [9]):

$$\rho_t + (\rho v)_x = 0, \tag{13a}$$

$$(\rho v)_t + (\rho v^2 + p)_x = 0, \tag{13b}$$

$$e_t + ((e + p)v)_x - \lambda T_{xx} = 0, \tag{13c}$$

where, as before, ρ is the density, v is the velocity, e the energy per unit volume,

$$e = \rho\epsilon + \tfrac{1}{2}\rho v^2, \tag{13d}$$

ϵ is the internal energy. In this section,

$$\epsilon = \frac{1}{\gamma - 1}\frac{p}{\rho} + Zq, \tag{13e}$$

where γ is a constant, $\gamma > 1$, q is the total available bonding energy ($q \leqslant 0$), and Z is a progress parameter for the reaction. $T = p/\rho$ is the temperature, and λ is the coefficient of heat conduction. Z is assumed to satisfy the rate equation

$$dZ/dt = -KZ, \tag{13f}$$

where

$$\begin{aligned} K &= 0 \qquad \text{if} \quad T = p/\rho \leqslant T_0, \\ K &= K_0 \qquad \text{if} \quad T = p/\rho > T_0. \end{aligned} \tag{13g}$$

T_0 is the ignition temperature and K_0 is the reaction rate. The equations of the preceding section are recovered if we set $\lambda = 0$, $q = \Delta$, and $K = \infty$. Equation (13f) is a reasonable prototype of the vastly more complex equations which described real chemical kinetics. Viscous effects have been omitted here; their inclusion in the present context has little effect and presents little difficulty. (Thus, we assume here a zero Prandtl number.)

In this paper, the approximation of the dissipation term is relegated to a separate fractional step, where it is to be handled by straightforward explicit finite differences. In view of (13e), and the perfect gas law $T = p/\rho$ (in appropriate units), this fractional step requires merely the approximation of

$$T_t = \lambda(\gamma - 1)T_{xx}. \tag{14}$$

The differencing of a heat conduction term alone introduces negligible numerical dissipation and the numerical diffusion converges to zero as λ converges to 0. More sophisticated techniques, in which the solution of the diffusion equation is also imbedded in the elementary Riemann-like solutions, and in which boundary layers are approximated by means of a random walk, are described elsewhere. They require a rather lengthly separate explanation. We are restricting ourselves here to the case of small, but not very small, λ, and note that our method does not drown the real effects of λ in numerically induced conduction.

All that remains to be done is to describe the solution of the Riemann problem for Eqs. (13) with $\lambda = 0$. This is done with the following *simplifying assumption:* whatever energy may be released during the time $k/2$ in a portion of the fluid is released instantaneously. This approximation is well in the spirit of our method (since it approximates Z be a piecewise constant function); it also has some physical justification [1].

SOLUTION OF A RIEMANN PROBLEM WITH CHEMISTRY

Our goal is to solve Eqs. (13) and the following data:

$$S_l(\rho = \rho_l, p = p_l, v = v_l, Z = Z_l) \qquad \text{for} \quad x \leqslant 0$$

and

$$S_r(\rho = \rho_r, p = p_r, v = v_r, Z = Z_r) \qquad \text{for} \quad x > 0$$

with $\lambda = 0$. The main result of this section (and of the paper) is the fact that a Riemann problem can be solved even when deflagrations and detonations are included along with shocks and rarefactions in the panoply of possible wave patterns. We begin by a partial review of the case $K_0 = 0$ (no chemistry; see [2, 6, 9]). The solution consists of a right state S_r, a left state S_l, a middle state $S_*(p = p_*, v = v_*)$, separated by waves which are either rarefactions or shocks. S_* is divided by the slip line $dx/dt = v_*$ into two parts with possibly differing values of ρ, ρ_{*r} to the right of the slip line and ρ_{*l} to its left, To determine v_* and p_* we proceed as follows: define the quantity

$$M_r = (p_r - p_*)/(v_r - v_*). \tag{15}$$

If the right wave is a shock,

$$M_r = -\rho_r(v_r - U_r) = -\rho_{*r}(v_* - U_r), \tag{16}$$

where U_r is the velocity of the right shock. From the Rankine–Hugoniot conditions one obtains

$$M_r = (p_r\rho_r)^{\frac{1}{2}}\, \phi_1(p^*/p_r), \qquad p_*/p_r \geqslant 1, \tag{17a}$$

where

$$\phi_1(\alpha) = \left(\frac{\gamma+1}{2}\,\alpha + \frac{\gamma-1}{2}\right)^{\frac{1}{2}}. \tag{17b}$$

If the right wave is a rarefaction, we find

$$M_r = (p_r\rho_r)^{\frac{1}{2}}\, \phi_2(p_*/p_r), \qquad p_*/p_r \leqslant 1, \tag{18a}$$

where

$$\phi_2(\alpha) = \frac{\gamma-1}{2\gamma^{\frac{1}{2}}}\,\frac{1-\alpha}{1-\alpha^{(\gamma-1)/2\gamma}}. \tag{18b}$$

Equation (18b) is derived through the use of the isentropic law $p\rho^{-\gamma} =$ constant and the constancy of the right Riemann invariant $\Gamma_r = 2(\gamma p/\rho)^{\frac{1}{2}}/(\gamma - 1) - v$. The function

$$\begin{aligned} \phi &= \phi_1(\alpha), \qquad \alpha \geqslant 1, \\ &= \phi_2(\alpha), \qquad \alpha \leqslant 1, \end{aligned} \tag{19}$$

is continuous at $\alpha = 1$, with $\phi(1) = \phi_1(1) = \phi_2(1) = \gamma^{\frac{1}{2}}$. Similarly, we define

$$M_l = (p_l - p_*)/(v_l - v_*); \tag{20}$$

if the left wave is a shock,

$$M_l = \rho_l(v_l - U_l) = \rho_{*l}(v_* - U_l), \tag{21}$$

where U_l is the velocity of the left shock. As on the right, $M_l = (p_l\rho_l)^{\frac{1}{2}}\ \phi(p_*/p_l)$, where $\phi(\alpha)$ is defined as in Eqs. (17) and (18). From (15) and (20),

$$p_* = (u_l - u_r + p_r/M_r + p_l/M_l)/((1/M_r) + (1/M_l)). \tag{22}$$

These considerations lead to the following iteration procedure: Pick a starting value p_*^0 (or values M_r^0, M_l^0), and then compute $p_*^{\nu+1}$, $M_r^{\nu+1}$, $M_l^{\nu+1}$, $q \geqslant 0$ using

$$\tilde{p}^\nu = (u_l - u_r + p_r/M_r^\nu + p_l/M_l^\nu)/((1/M_r^\nu) + (1/M_l^\nu)), \tag{23a}$$

$$p_*^{\nu+1} = \max(\epsilon, \tilde{p}^\nu), \tag{23b}$$

$$M_r^{\nu+1} = (p_r\rho_r)^{\frac{1}{2}}\ \phi(p_*^{\nu+1}/p_r), \tag{23c}$$

$$M_l^{\nu+1} = (p_l\rho_l)^{\frac{1}{2}}\ \phi(p_*^{\nu+1}/p_l). \tag{23d}$$

Equation (23b) is needed because there is no guarantee that in the course of iteration $\tilde{p}$ remains $\geqslant 0$. We usually set $\epsilon_1 = 10^{-6}$. The iteration is stopped when

$$\max(|M_r^{\nu+1} - M_r^\nu|, |M_l^{\nu+1} - M_l^\nu|) \leqslant \epsilon_2$$

(we usually picked $\epsilon_2 = 10^{-6}$); one then sets $M_r = M_r^{\nu+1}$, $M_l = M_l^{\nu+1}$, and $p_* = p_*^{\nu+1}$.

To start this procedure one needs initial values of either M_r and M_l (or p_*). The starting procedure suggested by Godunov appears to be ineffective, and better results were obtained by setting

$$p_*^0 = (p_r + p_l)/2.$$

We also ensured that the iteration was carried out at least twice, to avoid spurious convergence when $p_r = p_l$.

As noted by Godunov, the iteration may fail to converge in the presence of a strong rarefaction. This problem can be overcome by the following variant of Godunov's procedure: If the iteration has not converged after L iterations (we usually set $L = 20$), Eq. (12b) is replaced by

$$p_*^{\nu+1} = \alpha \max(\epsilon_1, \tilde{p}^\nu) + (1 - \alpha)p_*^\nu \tag{23b'}$$

with $\alpha = \alpha_1 = \frac{1}{2}$. If a further L iteration occur without convergence, we reset $\alpha_2 = \alpha_1/2$.

More generally, the program was written in such a way that if the iteration fails to converge after lL iterations (l integer), α is reset to

$$\alpha = \alpha_l = \alpha_{l-1}/2.$$

In practice, the cases $l > 2$ were never encountered. The number of iterations required oscillated between 2 and 10, except at a very few points.

Once p_*, M_r, M_l are known, we have

$$v_* = (p_l - p_r + M_r u_r + M_l u_l)/(M_r + M_l) \tag{24}$$

from the definitions of M_r and M_l.

Consider now the case $K_0 \neq 0$ ($\lambda = 0$); the right and left waves may now be CJ or strong detonations as well as shocks and rarefactions. The task at hand is to incorporate these possibilities into the solution of the Riemann problem.

The state S_r remains a constant state; v_r and ρ_r are fixed. The energy in S_r must change at constant volume (and thus can do no work). The change δZ_r in Z_r can be found by integrating Eqs. (13f), (13g), with $Z(0) = Z_r$ and $Z(k/2) = Z_r + \delta Z_r$, $\delta Z_r \leqslant 0$. The new pressure is

$$p_r + \delta p_r = p_r + (\gamma - 1)\,\delta Z\, q\rho_r, \tag{25}$$

see Eq. (7). We write $p_r^{\text{new}} = p_r + \delta p_r$, and drop the superscript new. (We shall need the old Z_r again and thus refrain from renaming $Z_r + \delta Z_r$.) Similarly, Z_l changes to $Z_l + \delta Z_l$, and a new p_l is found using the obvious analog of Eq. (25).

In S_* the values of Z differ from the values $Z_r + \delta Z_r$, $Z_l + \delta Z_l$. Let Z_{*l} be the value of Z to the left of the slip line and let Z_{*r} be the value of Z to the right of the slip line. The difference in energy of formation across the right wave is $\Delta_r = (Z_{*r} - (Z_r + \delta Z_r))q$, and across the left wave it is $\Delta_l = (Z_{*l} - (Z_l + \delta Z_l))q$. We iterate on the values Z_{*l}, Z_{*r}, Δ_r, Δ_l. In the first iteration, we set $Z_{*r} = Z_r + \delta Z_r$, $Z_{*l} = Z_l + \delta Z_l$, and thus $\Delta_r = \Delta_l = 0$, and carry out the iterations (23). When (23) has converged, a new pressure p_* is given, and new densities ρ_{*r}, ρ_{*l} can be found from Eqs. (16), (21), or the isentropic law. New temperatures $T_{*r} = p_*/\rho_{*r}$, $T_{*l} = p_*/\rho_{*l}$ are evaluated, Eqs. (13f), (13g) are solved, and new values Z_{*r}, Z_{*l}, Δ_r, Δ_l are found. If $\Delta_r \geqslant 0$ the right wave is either a shock or a rarefaction, and if $\Delta_r < 0$ the right wave is either a CJ detonation followed by a rarefaction or a strong detonation.

Let v_* be the velocity in S_*. Given Δ_r, Δ_l, we can find the velocities v_{CJr}, v_{CJl} behind possible CJ detonations on the right and left (Eq. (12)). If $v_* \leqslant c_{\text{CJr}}$ the right wave is a CJ detonation followed by rarefaction, and if $v_* > c_{\text{CJr}}$ the right wave is a strong detonation. The CJ state is unaffected by S_* (since it depends only on S_r) and as far as the Riemann solution is concerned it is a fixed state. If the right wave is a CJ detonation, we redefine M_r,

$$M_r = (p_{\text{CJ}} - p_*)/(v_{\text{CJ}} - v_*)$$

(p_{CJ} from Eq. (11c)). Then

$$M_{\mathrm{r}} = (\rho_{\mathrm{CJ}} p_{\mathrm{CJ}})^{\frac{1}{2}}\ \phi_2(p_*/p_{\mathrm{CJ}}), \qquad p_*/p_{\mathrm{CJ}} \leqslant 1. \tag{26}$$

If the right wave is a strong detonation, we find from (9)

$$M_{\mathrm{r}} = (p_{\mathrm{r}}\rho_{\mathrm{r}})^{\frac{1}{2}}\ \phi_3(\rho_{\mathrm{r}}\Delta_{\mathrm{r}}, p_{\mathrm{r}}, p_*),$$

where

$$(\phi_3(\alpha_1, \alpha_2, \alpha_3))^2 = \left(\frac{\gamma-1}{2} + \frac{\gamma+1}{2}\frac{\alpha_3}{\alpha_2}\right)\Big/\left(1 - \frac{(\gamma-1)\alpha_1}{\alpha_3 - \alpha_2}\right).$$

Similar expressions occur on the left. The iteration starts with M_{r}, M_{l} from the previous iteration, and written out in full, appears as follows:

$$\tilde{p}^{\nu} = (\tilde{v}_{\mathrm{l}} - \tilde{v}_{\mathrm{r}} + \tilde{p}_{\mathrm{r}}/M_{\mathrm{r}}^{\nu} + \tilde{p}_{\mathrm{l}}/M_{\mathrm{l}}^{\nu})/(1/M_{\mathrm{r}}^{\nu} + 1/M_{\mathrm{l}}^{\nu}), \qquad \nu \geqslant 0,$$

$$p_*^{\nu+1} = \max(\epsilon, \tilde{p}^{\nu}),$$

$$v_*^{\nu} = (\tilde{p}_{\mathrm{l}} - \tilde{p}_{\mathrm{r}} + M_{\mathrm{r}}^{\nu}\tilde{v}_{\mathrm{r}} + M_{\mathrm{l}}^{\nu}\tilde{v}_{\mathrm{l}})/(M_{\mathrm{r}}^{\nu} + M_{\mathrm{l}}^{\nu}),$$

where

$$\begin{aligned}
(\tilde{\rho}_{\mathrm{r}}, \tilde{p}_{\mathrm{r}}, \tilde{v}_{\mathrm{r}}) &= (\rho_{\mathrm{CJr}}, p_{\mathrm{CJr}}, v_{\mathrm{CJr}}) && \text{if right wave} = \text{CJ detonation,}\\
&= (\rho_{\mathrm{r}}, p_{\mathrm{r}}, v_{\mathrm{l}}) && \text{otherwise,}\\
(\tilde{\rho}_{\mathrm{l}}, \tilde{p}_{\mathrm{l}}, \tilde{v}_{\mathrm{l}}) &= (\rho_{\mathrm{CJl}}, p_{\mathrm{CJl}}, v_{\mathrm{CJl}}) && \text{if left wave} = \text{CJ detonation,}\\
&= (\rho_{\mathrm{l}}, p_{\mathrm{l}}, v_{\mathrm{l}}) && \text{otherwise,}\\
M_{\mathrm{r}}^{\nu+1} &= (p_{\mathrm{r}}\rho_{\mathrm{r}})^{\frac{1}{2}}\ \phi_3(\rho_{\mathrm{r}}\Delta_{\mathrm{r}}, p_{\mathrm{r}}, p_*^{\nu+1}) && \text{if right wave} = \text{strong detonation,}\\
&= (\tilde{p}_{\mathrm{r}}\tilde{\rho}_{\mathrm{r}})^{\frac{1}{2}}\ \phi(p_*^{\nu+1}/\tilde{p}_{\mathrm{r}}) && \text{otherwise,}\\
M_{\mathrm{l}}^{\nu+1} &= (p_{\mathrm{l}}\rho_{\mathrm{l}})^{\frac{1}{2}}\ \phi_3(\rho_{\mathrm{l}}\Delta_{\mathrm{l}}, p_{\mathrm{l}}, p_*^{\nu+1}) && \text{if left wave} = \text{strong detonation,}\\
&= (\tilde{p}_{\mathrm{l}}\tilde{\rho}_{\mathrm{l}})^{\frac{1}{2}}\ \phi(p_*^{\nu+1}/\tilde{p}_{\mathrm{l}}) && \text{otherwise.}
\end{aligned}$$

The complexity of this iteration is more apparent than real. It is stopped when it has converged, as before. New values of $Z_{*\mathrm{r}}$, $Z_{*\mathrm{l}}$, Δ_{r}, Δ_{l} are evaluated, and the iteration is repeated; this process is stopped when Δ_{r}, Δ_{l} change by less than some predetermined ϵ_3 over two successive iterations. It can be readily seen that with the present expression for the energy of formation, at most four iterations on Δ_{r}, Δ_{l} are ever needed.

Once S_* has been determined, the solution must be sampled. Let $P = (\theta h, k/2)$ be the sample point, and $\tilde{\rho} = \rho(P)$, $\tilde{p} = p(P)$, etc. Four basic cases are to be considered:

(A) P lies to the right of the slip line and the right wave is either a shock or a strong detonation;

(B) P lies to the right of the slip line and the right wave is either a rarefaction or a CJ detonation followed by a rarefaction;

(C) P lies to the left of the slip line and the left wave is either a shock or a strong detonation, and

(D) P lies to the left of the slip line and the left wave is either a rarefaction or a CJ detonation followed by a rarefaction.

Case A. The velocity U_r of the shock or the strong detonation can be found from the relationship

$$M_r = -\rho_r(v_r - U_r);$$

if P lies to the right of $dx/dt = U_r$ we have the sampled values $\tilde{\rho} = \rho_r$, $\tilde{p} = p_r$, $\tilde{v} = v_r$, $\tilde{Z} = Z_r + \delta Z_r$. If P lies to the left of $dx/dt = U_r$, we have $\tilde{\rho} = \rho_{*r}$, $\tilde{p} = p_*$, $\tilde{v} = v_*$, $\tilde{Z} = Z_{*r}$.

Case B. Consider first the case of a rarefaction wave. The rarefaction s bounded on the right by the line $dx/dt = v_r + c_r$, $c_r = (\gamma p_r/\rho_r)^{\frac{1}{2}}$, and on the left by $dx/dt = v_* + c_{*r}$, where c_* can be found by using the constancy of the Riemann invariant

$$\Gamma_r = 2c_*(\gamma - 1)^{-1} - v_* = 2c_r(\gamma - 1)^{-1} - v_r.$$

If P lies to the right of the rarefaction, $\tilde{\rho} = \rho_r$, $\tilde{p} = p_r$, $\tilde{v} = v_r$, $\tilde{Z} = Z_r + \delta Z_r$. If P lies to the left of the rarefaction, $\tilde{\rho} = \rho_{*r}$, $\tilde{p} = p_*$, $\tilde{v} = v_*$, $\tilde{Z} = Z_r + dZ_r$. If P lies inside the rarefaction, we equate the slope of the characteristic $dx/dt = v + c$ to the slope of the line through the origin and P, obtaining

$$\tilde{v} + \tilde{c} = 2\theta h/k;$$

the constancy of Γ_r, the isentropic law $p\rho^{-\gamma} =$ constant and the definition $c = (\gamma p/\rho)^{\frac{1}{2}}$ yield $\tilde{\rho}$, $\tilde{v}$, and $\tilde{p}$. $\tilde{Z} = Z_r + \delta Z_r$. If the wave is a CJ detonation, (ρ_r, p_r, v_r) are replaced by $(\rho_{CJr}, p_{CJr}, v_{CJr})$ in all formulas which describe the flow to the left of the detonation.

The cases C and D are mirror images of A and B and will not be described in full.

Numerical Results

We begin by presenting some results for detonation waves with very large K_0 ($K_0 = 1000$). These results verify the accuracy of the programming rather than the general validity of the method, since the solutions of the corresponding problems are an intrinsic part of the Riemann problem solution routine.

To obtain Table I, I started with a gas at rest, $\rho = 1$, $v = 0$, $p = 1$, and at $t = 0$ imposed impulsively on the left the boundary condition $v = V = 1$. I used $h = \frac{1}{7}$, $k/h = 2$, $K_0 = 1000$, $T_0 = 1.1$, $q = 1$, and $\gamma = 1.4$. The result is a perfect strong detonation.

TABLE I

Strong Detonation[a]

x	v	ρ	p	T	Z
0	1.	1.816	3.228	1.779	0.000
$\frac{1}{7}$	1.	1.816	3.228	1.779	0.000
$\frac{2}{7}$	1.	1.816	3.228	1.779	0.000
$\frac{3}{7}$	1.	1.816	3.228	1.779	0.000
$\frac{4}{7}$	0.	1.000	1.000	1.000	1.000
$\frac{5}{7}$	0.	1.000	1.000	1.000	1.000
$\frac{6}{7}$	0.	1.000	1.000	1.000	1.000
1	0.	1.000	1.000	1.000	1.000

[a] $h = \frac{1}{7}$, $k/h = .2$, $t = nk = .314$, $n = 11$, $K_0 = 1000$, $T_0 = 1.1$, $V = 1$, $q = 1$, $\gamma = 1.4$.

TABLE II

Flow Involving a Chapman–Jouguet Detonation[a]

x	v	ρ	p	T	Z	Γ_r
0	1.000	1.179	6.965	5.907	0.000	13.379
$\frac{1}{9}$	1.000	1.179	6.965	5.907	0.	13.379
$\frac{2}{9}$	1.000	1.179	6.965	5.907	0.	13.379
$\frac{3}{9}$	1.000	1.179	6.965	5.907	0.	13.379
$\frac{4}{9}$	1.186	1.257	7.621	6.061	0.	13.379
$\frac{5}{9}$	1.251	1.287	7.862	6.115	0.	13.379
$\frac{6}{9}$	1.524	1.410	8.952	6.346	0.	13.379
$\frac{7}{9}$	1.524	1.410	8.952	6.346	0.	13.379
$\frac{8}{9}$	1.623	1.457	9.373	6.430	0.	13.379
1	0.	1.000	1.000	1.000	1.000	5.916

[a] $h = \frac{1}{2}$, $k/h = .2$, $t = nk = .2$, $n = 9$, $K_0 = 1000$, $T_0 = 1.1$, $V = 1$, $q = 12$, $\gamma = 1.4$.

In Table II a Chapman–Jouguet detonation is exhibited. $h = \frac{1}{9}$, $k/h = 2$, $K_0 = 1000$, $T_0 = 1.1$, $q = 12$, and $\gamma = 1.4$. $m = 11$. The solution is exhibited at $t = 2$, $n = t/k = 9$, i.e., n is not a multiple of m and the solution is not at its most accurate. This can be seen from the presence of a fake constant state (for $x = \frac{6}{9}$ and $\frac{7}{9}$), which was discussed in the section about errors, and which is most likely to appear when n is not a multiple of m. The last column presents the right Riemann invariant Γ_r which is of course constant behind the CJ front. The chemical time scale is not resolved on the grid, and one should notice the small number of mesh points required to display sharp variations in all quantities.

TABLE III

Structure of an Exothermic Region with Finite Conduction and Reaction Rate[a]

x	v	ρ	p	T	Z
0	0.	.567	1.667	2.937	0.334
$\frac{1}{11}$	0.139	.650	1.781	2.739	0.614
$\frac{2}{11}$	0.261	.547	1.315	2.402	0.614
$\frac{3}{11}$	0.385	1.074	1.726	1.607	1.000
$\frac{4}{11}$	0.575	1.550	1.998	1.288	1.
$\frac{5}{11}$	0.544	1.519	1.800	1.185	1.
$\frac{6}{11}$	0.023	1.016	1.058	1.041	1.
$\frac{7}{11}$	0.002	1.001	1.003	1.002	1.
$\frac{8}{11}$	0.000	1.000	1.000	1.000	1.
$\frac{9}{11}$	0.	1.	1.	1.	1.
$\frac{10}{11}$	0.	1.	1.	1.	1.
1	0.	1.	1.	1.	1.

[a] $h = \frac{1}{11}$, $k/h = .35$, $t = nk = .273$, $n = 9$, $K_0 = 1$, $T_0 = 1.6$, $V = 0$, $q = 10$, $\gamma = 1.4$.

We now present some results for a problem whose solution is not programmed into the solution algorithm—a reaction zone with finite reaction rate. For $t < 0$ a gas at rest lies in $x \geqslant 0$, with $\rho = 1$, $p = 1$ ($v = 0$), and $Z = 1$; the left boundary is maintained at zero velocity, $V = 0$. At $t = 0$ the gas in the first cell to the left is raised to a temperature $T = 2$, (i.e., the pressure is increased to $p = 2$). The resulting deflagration wave is observed. It is known that the velocity of the wave is asymptotically proportional to $(\lambda K_0)^{\frac{1}{2}}$ (see e.g., [10, p. 99]); thus, the wave does not propagate unless $\lambda \neq 0$, as one can readily verify on the computer. This last justifies an earlier assertion to the effect that when $\lambda = 0$ the wave is indistinguishable from a slip line. The results in Table III were obtained with $h = \frac{1}{11}$, $k/h = 0.35$, $T_0 = 1.6$, $K_0 = 1$, $q = 10$, $\gamma = 1.4$, and $m = 11$. They are presented at $t = nk = 0.273$ ($n = 9$). One can clearly see the precursor shock, and the deflagration zone (characterized by $Z < 1$) in which the density and pressure decrease. The small number of mesh points should again be noticed; the wave has not yet settled to its asymptotic shape (and is thus quite interesting).

Conclusions

We have presented a numerical method capable of describing a complex gas flow with chemical reactions. The relative complexity of the method is balanced by economy in the representation of the solution. Generalization of the method to problems in more space dimensions is a straightforward application of the fractional step method presented in [2], and the inclusion of a more realistic chemical process presents no difficulties other than the standard difficulties of finding a plausible kinetic scheme and acceptable numerical values for the corresponding coefficients. The interesting and major difficulties in multidimensional problems arise when one attempts to take into account boundary layers and turbulence effects. In a forthcoming paper we shall show that boundary layer effects at least can be incorporated into our method in a natural and efficient way; once this has been explained, multidimensional results will be presented. It is expected that the method will be useful in those combustion problems where time dependence is an essential feature.

References

1. A. A. Borisov, *Acta Astronautica* **1** (1974), 909.
2. A. J. Chorin, *J. Comp. Phys.* **22** (1976), 517.
3. L. M. Cohen, J. M. Short, and A. K. Oppenheim, *Combustion and Flame* **24** (1975), 319.
4. R. Courant and K. O. Friedrichs, "Supersonic Flow and Shock Waves," Interscience, New York, 1948.
5. J. Glimm, *Comm. Pure Appl. Math.* **18** (1965), 697.
6. S. K. Godunov, *Mat. Sb.* **47** (1959), 271.
7. P. D. Lax, *SIAM Rev.* **11** (1969), 7.
8. A. Majda and S. Osher, to appear.
9. R. D. Richtmyer and K. W. Morton, "Finite Difference Methods for Initial Value Problems," Interscience, New York, 1967.
10. F. A. Williams, "Combustion Theory," Addison–Wesley, Reading, Mass., 1965.

Vortex Sheet Approximation of Boundary Layers

ALEXANDRE JOEL CHORIN*

Department of Mathematics and Lawrence Berkeley Laboratory, University of California, Berkeley, California, 94720

Received May 13, 1977; revised August 17, 1977

A grid free method for approximating incompressible boundary layers is introduced. The computational elements are segments of vortex sheets. The method is related to the earlier vortex method; simplicity is achieved at the cost of replacing the Navier-Stokes equations by the Prandtl boundary layer equations. A new method for generating vorticity at boundaries is also presented; it can be used with the earlier vortex method. The applications presented include (i) flat plate problems, and (ii) a flow problem in a model cylinder–piston assembly, where the new method is used near walls and an improved version of the random choice method is used in the interior. One of the attractive features of the new method is the ease with which it can be incorporated into hybrid algorithms.

INTRODUCTION

Some time ago we introduced a random vortex method for solving the Navier-Stokes equations [4]. The idea of the method was to approximate Euler's equations by analyzing the interaction of vortices, and then introduce the effects of viscosity by adding to the motion of the vortices an appropriate random component. This method has been further developed by, among others, Ashurst [1], Leonard [14, 15], Meng [19], Rogallo [20], and Shestakov [23, 24], and theoretical analyses have been carried out by Marsden *et al.* [17–8], among others. One attractive feature of the method is the fact that the tangential boundary condition is satisfied through vorticity creation, a procedure which mimics an essential physical phenomenon (see [2, 16, 25]).

That method has of course not solved all the outstanding problems of high Reynolds number flow. Some of the difficulties in its use have been: (i) the rate of convergence near boundaries has been slow, and as a result it is not always easy to ensure that the results obtained are independent of numerical parameters except possibly when points of separation can be determined a priori; (this point has been investitigated by Ashurst [1] and Rogallo [26]); (ii) the dependence of the method on an assumed structure of the vortices makes analysis difficult, in particular in the three dimensional case (see e.g. Leonard [14], [15]); (iii) in interior flow problems, the cost of the calculation can be substantial (see e.g. Ashurst [1]); Shestakov [23], [24] has derived a hybrid method which partly overcomes this difficulty. In the present paper, we present a new vorticity

* Partially supported by the Office of Naval Research under contract No. N00014-76-6-0316, and by the U.S. ERDA.

Reprinted with permission from *Journal of Computational Physics*, Vol. 27, No. 3, 1978.

generation method which should overcome problem (i) above, and introduce a related vortex method which solves the Prandtl boundary layer equations; in this method the vortex interaction is not singular, problem (ii) disappears, and the method can be used near boundaries in hybrid methods. A more general (but much more complicated) vortex method for the analysis of three dimensional turbulent boundary layers will be described elsewhere [7].

PRINCIPLE OF THE METHOD

The boundary layer equations can be written in the form (see e.g. Schlichting [21])

$$\partial_t \xi + (\mathbf{u} \cdot \nabla)\, \xi = \nu \partial_y{}^2 \xi, \tag{1a}$$

$$\xi = -\partial_y u, \tag{1b}$$

$$\partial_x u + \partial_y v = 0, \tag{1c}$$

where $\mathbf{u} = (u, v)$ is the velocity, u is tangential to the boundary and v is normal to the boundary, x is the spatial coordinate tangential to the boundary and y is the coordinate normal to the boundary, ξ is the vorticity and ν is the viscosity. We assume the wall is at $y = 0$ and the fluid fills out the half-space $y \geqslant 0$. The boundary conditions are

$$\mathbf{u} = 0 \text{ at } y = 0, \tag{2a}$$

$$u(x, y = \infty) = U_\infty(x). \tag{2b}$$

Additional conditions may be needed on the left and/or on the right. Equation (1c) can be integrated in the form

$$v(y) = -\partial_x \int_0^y u(x, z)\, dz \tag{3a}$$

and Eq. (1b) yields

$$u(x, y) = U_\infty - \int_y^\infty \xi(x, z)\, dz. \tag{3b}$$

It can be readily seen that if ξ is known, (3a) and (3b) yield u and v.

Consider a collection of N segments S_i of vortex sheets, of intensities ξ_i, $i = 1,\ldots, N$ (i.e., segments of a straight line such that u on one side of S_i and u on the other side of S_i differ by ξ_i). S_i is parallel to the x axis, of length h and center $\mathbf{x}_i = (x_i, y_i)$. The x component of the velocity of S_i due to the presence of the other segments can be found from (3b), which yields approximately

$$u_i = U_\infty(x_i) - \tfrac{1}{2}\xi_i - \sum_j \xi_j d_j \tag{4a}$$

where

$$d_j = 1 - (| x_i - x_j |/h) \tag{4b}$$

and the sum $\sum$ is over all S_j such that

$$y_j > y_i \text{ and } | x_i - x_j | < h \tag{4c}$$

(i.e., $0 \leqslant d_j \leqslant 1$).

The vertical velocity v_i of S_i can be approximated from (3a) by

$$v_i = -(I_1 - I_2)/h \tag{5a}$$

where I_1 and I_2 approximate respectively $\int_0^{y_i} u(x + h/2, y)\, dy$ and $\int_0^{y_i} u(x - h/2, y)\, dy$, and can be taken as

$$I_1 = U_\infty(x_i + h/2)\, y_i - \sum_{j^+} \xi_j d^+{}_j y^*{}_j\,, \tag{5b}$$

$$I_2 = U_\infty(x_i - h/2)\, y_i - \sum_{j^-} \xi_j d^-{}_j y^*{}_j\,, \tag{5c}$$

where

$$d^+{}_j = 1 - | x_i + h/2 - x_j |/h, \tag{5d}$$

$$d^-{}_j = 1 - | x_i - h/2 - x_j |/h, \tag{5e}$$

$$y^*{}_j = \min(y_i\,, y_j), \tag{5f}$$

the sum $\sum_+$ is over all S_j such that $0 \leqslant d^+{}_j \leqslant 1$ and the sum $\sum_-$ is over all S_j such that $0 \leqslant d^-{}_j \leqslant 1$. Note that the total number of interactions between vortex sheets is small, in particular in comparison with what happens when point vortex interactions are taken into account; one can use sorting algorithms to minimize the number of decisions involved in carrying out the several summations.

Thus, the motion of vorticity described by the equations

$$\partial_t \xi + (\mathbf{u} \cdot \boldsymbol{\nabla})\, \xi = 0,$$

$$\partial_x u + \partial_y v = 0,$$

$$\xi = -\partial_y u$$

can be approximated by

$$x_i^{n+1} = x_i^{\,n} + k u_i\,,$$

$$y_i^{n+1} = y_i^{\,n} + k v_i\,,$$

where k is a time step, and $x_i^n \equiv x_i(nk)$, $y_i^n \equiv y_i(nk)$. The effect of viscosity can then be included by adding to the deterministic formula for y_i^{n+1} a random variable η_i drawn from a gaussian distribution with mean 0 and variance $2\nu k$; this yields the algorithm

$$x_i^{n+1} = x_i^n + ku_i , \tag{6a}$$

$$y_i^{n+1} = y_i^n + kv_i + \eta_i . \tag{6b}$$

The several values of η_i are independent, u_i is given by (4) and v_i by (5). The boundary conditions $u = U_\infty$ at $y = \infty$ and $v = 0$ at $y = 0$ are automatically satisfied; the boundary condition $u = 0$ at $y = 0$ will be satisfied through a vorticity creation operation described in the next section. The statistical error in Eq. (6) can be reduced by a tagging method which will also be described below. Note that no grid is introduced; there is no lower bound to the thickness of the boundary layer which can be resolved, and no differencing occurs across the layer. Furthermore, the solution is computed in the (x, y) plane, without a change of variables, and thus it should be easy to match the computed boundary layer solution with an inviscid solution outside the layer.

Vorticity Creation

In [4] we proposed the following algorithm for satisfying the tangential boundary condition on $\mathbf{u}$: Let $\mathbf{u}_0$ be the flow satisfying the equation of motion and the boundary condition $v_0 = 0$. If at the wall $u_0 \neq 0$, the effect of viscosity will be to create a thin boundary layer near the wall; the total vorticity in the layer per unit length of the wall is

$$\int_{\text{wall}}^{\text{interior}} \xi \, dy = \int_{\text{wall}}^{\text{interior}} \frac{\partial u}{\partial y} \, dy = u_0$$

i.e., one has to create a vortex sheet of strength u_0 per unit length of the wall; this vortex sheet is then broken up into elements and allowed to participate in the subsequent motion of the fluid. The vorticity elements which cross the wall are lost; their vorticity will of course be recreated at the next step. This construction was offered in [4] on heuristic and physical grounds.

To understand the nature of the approximations made, it is adequate to consider the diffusive part of the equation, i.e., the diffusion equation $\partial_t \xi = \nu \partial_y^2 \xi$ with the boundary $u = 0$. The gaussian random variable provides an approximate solution of the whole space heat equation (since the Green's function of the heat equation in the absence of boundaries is a gaussian function). The subsequent deletion of the vortices which cross the boundary and the creation of a vortex sheet of intensity u serve to project the solution of the whole space heat equation on the subspace of functions which vanish outside the domain of integration. This formulation is due to Marsden and McCracken; its convergence as $k \to 0$ in the case of linear equations such as the heat equation follows from the work of Kato [11, 12] (see [8] for a review). It has, however, been observed, computationally by Ashurst [1], Rogallo [20] and the author, and theoretically in [8], that the rate of convergence near the wall as $k \to 0$ is slow, in particular since the boundary condition $u = 0$ on the wall is satisfied only in the limit. We therefore introduce an alternative to the earlier approximation in which the boundary condition is satisfied exactly except possibly at a finite set of values of t. The velocity field is extended across the wall at the beginning of each time step by the anti-symmetry $u(x, -y) = -u(x, y)$, where the wall is assumed to be at $y = 0$. As a result, $\xi(x, -y) = \xi(x, y)$ for $y \neq 0$, and a vortex sheet appears at $y = 0$ if the tangential velocity does not vanish at the wall. This anti-symmetry replaces the vorticity creation operation used in earlier work. The whole space diffusion equation is then solved by a random walk, for a time k. using as initial data the extended solution. Algorithmically, this is equivalent to (i) creating a vortex sheet of strength $2u_0$ per unit length of the wall, and (ii) bouncing those vortices which cross the wall back into the fluid, i.e., if at the end of a time step a vortex finds itself at (x_i, y_i), $y_i < 0$, it is returned to $(x_i, -y_i)$. For some analysis, see [8].

Thus, we take points $Q_1, \ldots, Q_m$ at the wall, such that the distances $\overline{Q_1Q_2}$, $\overline{Q_2Q_3}$ equal h. At each point Q_i we evaluate the tangential velocity u_0, using the obvious specialization of Eqs. (4). We imagine then a vortex sheet of strength $2u_0$ at Q_i. In order to have a reasonable approximation of the diffusion equation at a later time, we create at Q_i not a simple vortex sheet, but some number l of sheets such that the intensity of each is less in absolute value than a predetermined $\xi_{\max}$. At the next step, these sheets will behave according to the laws (6). Some obvious programming precautions must be taken: the vortex sheets which have just been created and are taking their first random step may jump out of the domain of integration; these should be lost and not bounced back (or else the wall symmetry will be violated). One must also ensure that the term $\frac{1}{2}\xi_i$ in the formula (4a) does not add an unnecessary horizontal component to the motion of the newly created sheets.

A substantial reduction in the statistical error can be made by observing that in Eq. (1) diffusion takes place only in the y direction; thus the numbers η_i used in (6) need be independent of each other only when they are used with vortex segments whose centers lie in a narrow strip perpendicular to the wall. This fact can be used in the following way: As vortices are created, they are assigned integer tags, m_i being the tag assigned to the ith vortex sheet element. At each time step, a tag not used before is chosen and assigned to one vortex element at each boundary point Q_j at which at least one element is created. A second tag is then chosen, and assigned to one element at each point where at least two elements are created, etc. The effect of the tagging is to piece together the elements created at the several boundary points into coherent vortex sheets, with the elements of each sheet identified by a common tag. When the random numbers η_i are chosen for use in (6), all elements with the same tag are assigned the same η. This is the variance reduction procedure. In parallel flows, its effect is to make the sums in (5) identically zero (and thus reduce their variance to zero).

Flow Past a Semi-infinite Flat Plate

Consider a semi-infinite flat plate placed on the positive x axis, with a fluid of density 1 occupying the half plane $y > 0$. At time $t = 0$ the fluid is impulsively set in motion with velocity $U_\infty = 1$. We shall apply our method to the analysis of this problem, with the aim of comparing the results with the well known solution (see, e.g., Schlichting [21]).

The leading edge singularity presents no difficulty. One fairly minor detail requires some attention: we are going to compute over a finite length of the plate, say for $0 \leqslant x \leqslant a$. From Eqs. (1) it follows that no boundary condition need, or indeed may, be imposed at $x = a$, since the flow of information will be to the right only. However, formula (5a) is essentially a centered difference approximation to $\partial_x \int u\, dy$, and may give rise to a spurious flow of information to the left. This is easily corrected by removing all vortex sheets which cross $x = a$ and by not allowing those sheets whose centers lie between a and $a - 2h$ to have any motion in the y direction — they are thus merely convected downstream without disturbing the flow to their left.

The numerical parameters at our disposal are h, k, and $\xi_{\max}$. The method is unconditionally stable, and h, k are constrained only by an accuracy requirement $uk \leqslant O(h)$. Convergence should occur as $h, k, \xi_{\max}$ all tend to zero. As these parameters decrease, the number N of sheets in the calculation increases, the amount of labor increases, but both the differencing error in (5) and the statistical error decrease.

The calculations were pursued until a steady state had been reached and maintained for a while. In a steady state, the drag D on the portion of the plate between O and a point X can be evaluated from the momentum defect formula ([21, p. 161])

$$D = \int_0^\infty u(U_\infty - u)\, dy, \qquad u = u(X, y).$$

The integral can be evaluated as follows: Consider all the vortex sheets S_i, $i = 1, 2, \ldots, M$ whose centers satisfy $| x_i - X | < h$. Assume that they are numbered in such a way that $y_1 \leqslant y_2 \leqslant y_3 \leqslant \cdots \leqslant y_{12}$. Then we have approximately

$$D = \sum_{i=1}^{M} u_i(U_\infty - u_i)\, \Delta y_i\,,$$

where, as before,

$$u_i = U_\infty(X) - \tfrac{1}{2}\xi_i - \sum_{j=i+1}^{M} \xi_j d_j\,,$$

with $d_j = | x_j - X |/h$, $\Delta y_i = y_i - y_{i-1}$, $y_0 = 0$.

Define the streamwise Reynolds number

$$R = U_\infty X/\nu;$$

to first order in $R^{-1/2}$ we have from boundary layer theory

$$D = 0.664/R^{1/2}. \tag{7}$$

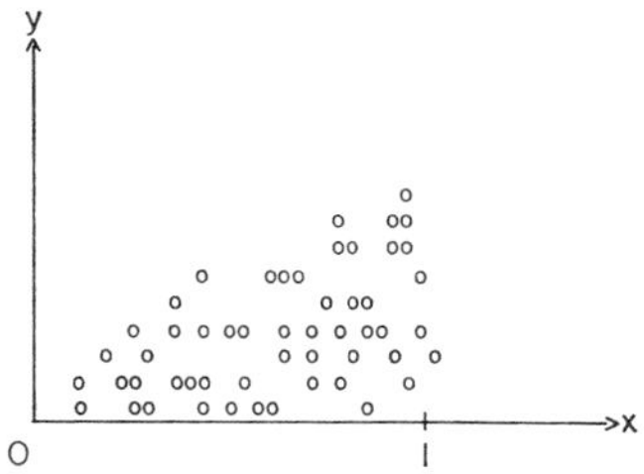

FIG. 1. Vortex sheets over a flat plate.

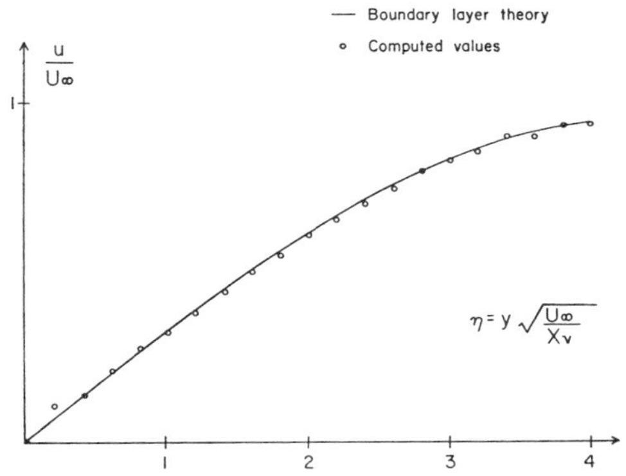

FIG. 2. Horizontal velocity in Blasius flow.

In Fig. 1 we display a typical vorticity configuration: an O corresponds to the center of a vortex sheet. This configuration was obtained with $k = 0.2$, $h = 0.2$, $\xi_{\max} = 0.1$, $\nu = 10^{-6}$, at $t = 5.0$.

We found experimentally that for $k \leqslant 0.2$, $h \leqslant 0.2$, $\xi_{\max} \leqslant 0.1$ the statistical error dominated all others; this error decreases rather slowly as the number of vortex sheets increases, but will not be particularly troublesome in later applications (see below). One method for reducing the statistical error in the steady state is to average the solution over a number of time steps (see [23, 24]). In Fig. 2 we display the velocity profile averaged over 20 steps with $\nu = 10^{-6}$, $k = 0.2$, $h = 0.2$, $\xi_{\max} = 0.1$, $8 \leqslant t \leqslant 12$, compared with the analytic boundary layer solution.

The drag computed at $\nu = 10^{-6}$, averaged over 20 steps, is 6.69×10^{-2}, compared with the value 6.64×10^{-2} obtained from (7). If one considers the successive values of D at the several time steps to be successive estimates of D, then the standard deviation of the computed answer is 0.4×10^{-2}. At $\nu = 10^{-4}$, the computed value of D is 0.669, with standard deviation .04, compared to the value $D = 0.664$ obtained from (7). In all our calculations, the averages of the computed D converged to the mean value much faster than one would have expected from the estimates of the standard deviation. No explanation is offered, and we do not know how general this effect may be. The typical number of vortex sheets in these runs is 100, and a typical running time is 20 seconds on the CDC 6400 computer at Berkeley.

I also ran some problems where U_∞ was not constant but had the form

$$U_\infty = 1 - A \sin \pi x.$$

TABLE 1

Separation and Reattachment

	x					
η	0.1	0.2	0.3	0.4	0.5	0.6
0.04	0.34	0.15	0.00	−0.12	−0.04	0.04
0.08	0.38	0.17	0.02	−0.10	−0.03	0.05
0.12	0.39	0.18	0.03	−0.09	−0.01	0.06
0.16	0.39	0.19	0.04	−0.07	0.00	0.09
0.20	0.41	0.21	0.05	−0.06	0.01	0.09
0.24	0.42	0.22	0.06	−0.05	0.02	0.10
0.28	0.44	0.23	0.07	−0.05	0.02	0.11
0.32	0.44	0.24	0.08	−0.03	0.04	0.11
0.36	0.45	0.24	0.10	−0.02	0.03	0.12
0.40	0.46	0.26	0.10	−0.02	0.04	0.13
0.44	0.46	0.27	0.11	−0.00	0.05	0.13
0.48	0.46	0.28	0.12	0.01	0.06	0.12
0.52	0.47	0.28	0.14	0.03	0.07	0.12

For large enouch A one expects separation, reattachment and a recirculation bubble. However, in a direct method (as opposed to an inverse method [13]) one expects the singularities at separation and reattachment to taint the solution (see [22, 9]). In Table I we display some values of u obtained with $k = .1$, $h = .1$, $\xi_{\max} = 1$, $\nu = 10^{-6}$, $A = .2$, averaged over 20 steps between $t = 8$ and $t = 12$. $\eta = y/(X\nu)^{1/2}$ is the usual similarity variable. The negative values of u represent the recirculation region. A steady state was never achieved; the validity of the solution is unclear and it is in fact doubtful. No connection with the inverse method of Klineberg and Steger [13] was established. The effect of the singularities is widely believed to be removable by coupling the boundary layer calculation to the outer calculation; a method for doing this for another problem is described in the following section.

A Hybrid Algorithm Involving the Random Choice Method

We now present a hybrid algorithm in which the method described above is used near the boundaries while a different method is used in the interior of the domain. The two components of the algorithm are coupled, with the vortex sheet method serving as vorticity source for the interior method. An earlier hybrid method was presented by Shestakov [23, 24]; in Shestakov's work, a vortex blob method was used near the walls, and a difference method was used in the interior, with a coupling based on a careful use of spline interpolation. A hybrid method based on the use of vortex sheets near walls and vortex monopoles in the interior will be presented elsewhere [7].

Here we use as an interior method a version of the random choice method for compressible flow [5, 6]. Thus, not only do we use different methods in the interior and near walls, but we also make different assumptions about compressibility: We have viscous incompressible flow near the walls and inviscid compressible flow in the interior. There are two sets of reasons for doing this:

(a) Difficulties with interior viscosity. One may well believe that the numerical viscosity associated with finite difference or finite element methods has little effect as long as one stays away from walls, but it is not clear what "staying away from walls" should mean. The interior method must reach quite close to the walls, and earlier numerical experiments [23] indicate that unless the interior viscosity is tightly controlled, e.g., through the use of a very fine grid, the results may be substantially in error. The random choice method has effectively no numerical viscosity and is available for use. Since all we want to do is demonstrate how the sheet method can be coupled to an interior method, the random choice method is acceptable, as long as the Mach number near the walls is reasonably small.

(b) Ulterior motives. The methods of this paper will be used on the analysis of reacting gas flow, and in that context it is believed that the particular mixture of methods we use here will be most appropriate.

The most important problem is to find a reasonable way for coupling the interior and the boundary. If it is known in advance that the boundary layer will not separate, this is trivial, since all one has to do is use tangential velocities from the interior as velocities at infinity for the boundary. In interesting cases it is, however, essential that the layer act on the interior as well, since it may have a crucial impact on the interior flow, and since some boundary–interior interaction is needed to counteract the separation singularity. In the examples described in the following section we proceeded as follows: used the tangential velocity at the wall of the interior calculation as velocity at infinity for the boundary layer calculation, and impressed upon the interior calculation the velocity normal to the wall induced by the boundary layer calculation.

This last normal velocity was computed as follows: let P be a point at the wall, with coordinates $((l + \frac{1}{2})\,\Delta, 0)$, where l is an integer and Δ is the grid size in the interior. The momentum lost due to the boundary layer above P can be approximated by

$$U_{l+1/2} = \sum \xi_j d_j y_j$$

where $d_j = | x_j - (l + \frac{1}{2}) \Delta |/h$, (x_j, y_j) is the center of the vortex sheet S_j with vorticity ξ_j, and the sum is over all S_j such that $0 \leqslant d_j \leqslant 1$ (see Eqs. (5)). Then the normal velocity at $x = l\Delta$ is approximately $(U_{l+1/2} - U_{l-1/2})/\Delta$. This velocity is imposed on the interior calculation at the boundary.

The programming details of the joint vortex sheet-random choice calculation require a somewhat lengthy explanation, mostly because of the relative complexity of the random choice program. The equations solved in the interior are the usual Euler equations. As described in [5], one full step of the random choice method for these equations consists of four quarter steps of length $k/2$. Let $\mathbf{V}_{i,j}^n = \mathbf{V}(i\Delta, j\Delta, nk)$ denote the solution vector. At the beginning of the step we have $\mathbf{V}_{i,j}^n$ for i, j integers. In the first quarter step we compute $\mathbf{V}_{i+1/2,j}^{n,1/4}$, in the second quarter step we compute $\mathbf{V}_{i+1/2,j+1/2}^{n,1/2}$, in the third quarter step we compute $\mathbf{V}_{i,j+1/2}^{n,3/4}$, and in the last quarter step we compute $\mathbf{V}_{i,j}^{n+1}$. To obtain one new value for the vector $\mathbf{V}$ at a point one solves a Riemann problem, which is them sampled. The sampling strategies have been described in some detail in [6]; they involve "random" numbers θ. A Riemann problem is an initial value problem for the equations of motion in which the initial data are discontinuous. Its solution contains a slip line; i.e., a line which divides the fluid initially to the left of the discontinuity from the fluid initially to the right of the discontinuity.

Near the boundary, symmetry conditions can be used to formulate the appropriate Riemann problems. In the program used here, which is a refinement of the earlier program [5], the physical domain is not always fixed with respect to the computational grid. All points are identified by an integer tag q, with $q = 1$ for points in the interior of the domain and $q = 0$ for points outside the domain. q is treated as a passive quantity and propagates as part of the calculation, depending on the relative position of the slip line and the sampling point. If $q = 0$ for both initial states in the Riemann problem no calculation need be carried out. If $q = 1$ for both states we have an interior point, and if we have two distinct values of q the boundary symmetry conditions are applied. As already partly described in [6], if the "random" numbers θ are picked so that the first two are $\geqslant 0$, the next two $\leqslant 0$, etc., and if the bottom and left boundaries coincide with lines $x = I\Delta$, $y = J\Delta$, I, J integers, while the top and right boundaries coincide with lines $x = (I' + \frac{1}{2}) \Delta$, $y = (J' + \frac{1}{2}) \Delta$, I', J' integers, then stationary boundaries remain stationary on the grid. If the boundaries are chosen as we have just described, then one boundary layer calculation step must be made every two interior quarter steps, and the conditions at infinity for the boundary layer calculation can be updated and the normal velocity imposed on the interior only once every four quarter steps (i.e., once per whole interior step), the updating occurring whenever appropriate boundary data from the interior calculation are available. It should be obvious that the fact that the random choice method does not smear out vortex sheets immediately is helpful to the success of the method.

The accuracy of the method has not yet been discussed. Clearly, our matching procedure is based on the assumption that the boundary layer thickness is at most comparable with Δ, i.e., $\Delta \geqslant O(R^{-1/2})$, where R is a Reynolds number based on an interior length scale and velocity. The accuracy of the interior Glimm method is at best $O(\Delta)$ (see [4]). Thus, the over-all accuracy is at best $O(\Delta) + O(R^{-1/2})$. This is not a surprising estimate (see, e.g., [4] for a discussion), and if it can be shown to be realistic and to hold uniformly in $R^{-1/2}$, it would represent a substantial achievement. There are of course no problems with stability, since each component of the hybrid method is unconditionally stable.

Application to Two Dimensional Flow behind a Piston

We now present an application of the preceding algorithm, an application for which the random choice interior method is well suited. We do so with words of caution. The belief that our method can handle properly the separation of a boundary layer is based more on hope than on hard analysis. The accuracy of the results is difficult to gauge through the inevitable statistical error. There are no reliable data for comparison. The best that can be said is that the results are plausible, consistent with earlier work on similar problems (see, e.g., Bernard [3]), and consistent also with the belief that the effective diffusion of the scheme equals the nominal diffusion (i.e., that the computational results correspond to the Reynolds number explicitly imposed on the calculation and not to a numerical Reynolds number intrinsic to the method).

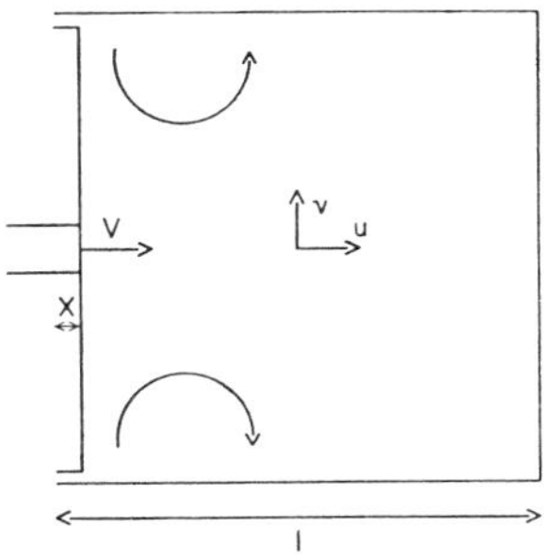

FIG. 3. Piston-cylinder flow configuration.

The flow configuration is shown in Fig. 3. A piston is pushed with velocity V into a chamber filled with gas. The initial density of the gas is $\rho = 1$, the initial pressure is $p = 1$, and the gas is initially at rest. The gas is assumed to be perfect, i.e., the internal energy is given by $\epsilon = (p/\rho)/(\gamma - 1)$, where $\gamma = 1.4$. The sound speed is $c = (\gamma p/\rho)^{1/2}$. The viscosity ν is measured in units in which $(p/\rho)^{1/2} = 1$ and the initial length of the chamber is 1. The width of the chamber is 1. Thus, the Reynolds number based on the velocity at infinity seen by the boundary layer and on the length of the chamber is

$R_V = V/\nu$. Care is taken to ensure that the Mach number $V/c \ll 1$. $V = 0$ for $t \leqslant 0$, and assumes a constant value for $t > 0$. The displacement of the piston is $X = Vt$.

In the absence of viscosity we would have a shock wave propagating into the gas, reflected at the far end, and then bouncing back and forth between the piston and the back wall. The random choice method would compute this flow with infinite resolution (see the analysis in [6]). Call this flow $\mathbf{u}_0 = (u_0, 0)$.

The effect of $\nu \neq 0$ is to superimpose on u_0 a rotational flow with the general pattern depicted in Fig. 3. The boundary layers on the top and bottom are slowed down and deflect some fluid at the piston towards the interior of the domain (see, e.g., [3]). We exhibit a calculation made with $\nu = 10^{-3}$. This relatively high value of ν is picked because the rotational effect we wish to exhibit decreases with ν. It is clear that as ν decreases our method does not break down. This ν is as large as we could pick and still observe the constraint $\Delta = O(\nu t)$. The results below must be considered while keeping in mind (i) the built-in fluctuations of the random choice method, (ii) the fact that the edge of the calculation is the edge of the boundary layer and not the boundary of the domain, and (iii) the coarseness of the interior grid.

The following parameters were used: In the interior, $\Delta = 1/13$, $k/\Delta = 0.6$, $m_1 = 7$, $m_2 = 3$ (these integers are used in the generation of the numbers θ which define the algorithm; see [5]). In the boundary layer, $h = 2\Delta = 2/13$ and $\xi_{\max} = V/5$. The

TABLE II

Horizontal Velocity behind a Piston ($t = 1.846$)

	x								
y	5/13	6/13	7/13	8/13	9/13	10/13	11/13	12/13	1
0	0.20	0.12	0.13	0.21	0.18	0.20	0.20	0.22	0.03
1/13	0.20	0.20	0.26	0.22	0.17	0.22	0.18	0.17	−0.01
2/13	0.20	0.18	0.18	0.27	0.24	0.20	0.20	0.21	0.03
3/13	0.20	0.25	0.22	0.21	0.23	0.20	0.19	0.17	0.01
4/13	0.20	0.22	0.20	0.20	0.19	0.19	0.19	0.17	0.00
5/13	.20	0.20	0.20	0.22	0.22	0.20	0.19	0.20	0.02
6/13	0.20	0.22	0.23	0.20	0.18	0.19	0.19	0.19	0.02
7/13	0.20	0.22	0.22	0.21	0.19	0.19	0.19	0.19	0.00
8/13	0.20	0.19	0.20	0.19	0.20	0.19	0.19	0.21	0.00
9/13	0.20	0.21	0.19	0.18	0.18	0.18	0.18	0.19	−0.00
10/13	0.20	0.23	0.24	0.22	0.23	0.21	0.17	0.17	0.00
11/13	0.20	0.17	0.21	0.25	0.18	0.16	0.19	0.18	−0.05
12/13	0.20	0.19	0.23	0.19	0.27	0.25	0.19	0.13	−0.03
1	0.20	0.21	0.16	0.21	0.21	0.22	0.22	0.04	−0.03

TABLE III

Vertical Velocity behind a Piston ($t = 1.846$)

	x								
y	5/13	6/13	7/13	8/13	9/13	10/13	11/13	12/13	1
0	0.04	−0.04	0.01	−0.02	0.03	0.04	0.01	0.01	−0.01
1/13	−0.12	−0.09	−0.00	0.03	−0.01	0.03	0.06	0.04	0.01
2/13	0.02	−0.04	−0.03	0.02	.01	0.00	0.01	0.03	0.01
3/13	0.02	0.00	0.01	0.01	−0.02	0.02	0.02	−0.00	0.00
4/13	−0.04	−0.04	−0.04	−0.03	−0.01	−0.00	0.01	0.01	−0.01
5/13	0.03	0.03	0.04	0.03	0.02	0.00	−0.01	−0.04	−0.03
6/13	−0.01	−0.03	−0.04	−0.03	−0.01	0.01	0.01	0.01	0.02
7/13	−0.01	−0.01	−0.01	0.01	0.03	0.02	0.02	0.02	0.00
8/13	0.01	0.01	−0.01	0.00	0.01	0.01	0.01	0.01	0.02
9/13	0.04	0.01	−0.01	0.02	0.00	0.02	−0.00	0.04	0.03
10/13	0.01	−0.02	−0.01	−0.01	0.02	−0.03	−0.02	−0.02	−0.02
11/13	0.08	0.01	−0.03	0.02	0.00	−0.02	−0.01	−0.03	−0.01
12/13	0.12	−0.01	−0.07	0.03	−0.01	−0.05	−0.07	−0.07	0.01
1	0.13	−0.03	−0.09	0.02	−0.04	−0.06	−0.03	−0.01	−0.02

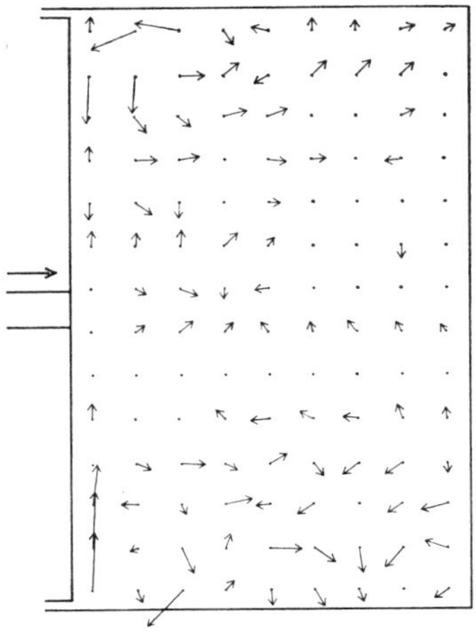

FIG. 4. Velocity in the piston-cylinder flow.

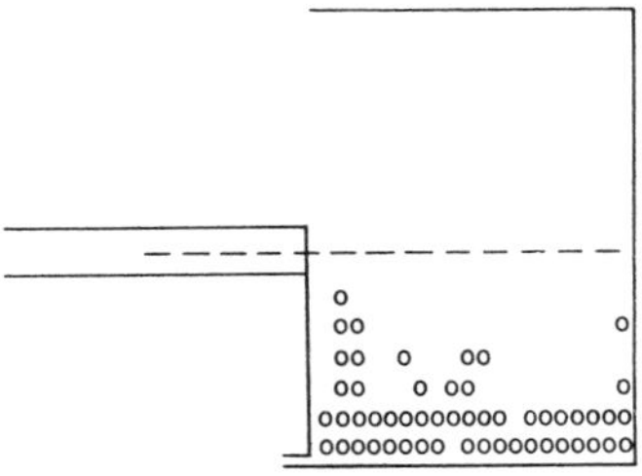

FIG. 5. Vorticity in the boundary layer of the piston-cylinder flow.

results displayed are at $t = 40k = 1.846$, when X = displacement of the piston = 0.3692. In Tables II and III we display the values of the horizontal and vertical velocity fields. In Fig. 4 we plot the vectors $(u - u_0, v)$, i.e., the difference between the flow with $\nu = 0$ and the flow with $\nu \neq 0$. The correct rotational behavior can be observed. In Fig. 5 we display the positions of the vortex centers in the lower half of the domain. At this t, there are 531 vortex sheets in the calculations, and the total computing time has been 9 minutes on a CDC 6400 computer. It must be pointed out that the boundary layer thickness is $O((\nu t)^{1/2})$; i.e., it varies from 0 to $O(\Delta)$, and that we are considering effects induced by the internal mechanics of the boundary layer, which would normally require a fine grid for adequate resolution.

CONCLUSION

We have presented a grid free method for studying boundary layers. The two main features of this method are: (i) the use of vortex sheet segments as computational elements, and (ii) a new method for generating vorticity at walls. It is expected that this algorithm will be mainly useful as a component of hybrid methods, and an example of such use has been given.

One can see that an algorithm based on non-rotating vortex sheets cannot reproduce the effects characteristic of turbulent boundary layers (see, e.g., [7]). Turbulence effects can conceivably be taken into account by replacing the molecular viscosity which determines the variance of the random variable η by an eddy viscosity. However, in later work we expect to use our present algorithm as a vorticity generation method for a hybrid method, in which the main part of the calculation will be carried out through the use of vortex elements of more elaborate structure; the sheets will be effectively relegated to the viscous sublayer.

It is obvious that a price must be paid for the removal of numerically induced viscosity in our method, and this price is statistical error. It is hoped that there will be a substantial number of applications in which such price is worth paying. It is also obvious that the present method generalizes trivially to three dimensional flows.

Note. The programs used to obtain the results above are available from the author.

Acknowledgment

I would like to thank Dr. A. Leonard and Professors M. McCracken and J. Marsden for many helpful discussions and comments.

References

1. W. Ashurst, Numerical simulation of turbulent mixing layer dynamics, to appear.
2. G. K. Batchelor, "An Introduction to Fluid Mechanics," Cambridge Univ. Press, London/New York, 1967.
3. P. S. Bernard, Ph.D. Thesis, University of California at Berkeley, 1977.
4. A. J. Chorin, *J. Fluid Mech.* **57** (1973), 485.
5. A. J. Chorin, *J. Computational Phys.* **22** (1976), 517.
6. A. J. Chorin, *J. Computational Phys.* **25** (1977), 253.
7. A. J. Chorin, Numerical study of turbulent boundary layer structure, to appear.
8. A. J. Chorin, T. J. R. Hughes, M. T. McCracken, and J. E. Marsden, Product formulas and numerical algorithms, *Comm. Pure Applied Math.* (1978), in press.
9. H. Dwyer and F. S. Sherman, manuscript (1976).
10. J. Glimm, *Comm. Pure Appl. Math.* **18** (1965), 697.
11. T. Kato, Trotter's product formula for an arbitrary pair of self-adjoint contraction semi-groups, to appear.
12. T. Kato, manuscript, 1976.
13. J. M. Klineberg and J. L. Steger, *AIAA Paper*, 74–94, 1974.
14. A. Leonard, *in* "Proceedings of the Fourth International Conference on Numerical Methods in Fluid Dynamics," Springer, New York, 1975.
15. A. Leonard, *in* "Proceedings of the Fifth International Conference on Numerical Methods in Fluid Dynamics," Springer, New York, 1977.
16. M. J. Lighthill, *in* "Laminar Boundary Layers" (L. Rosenhead, Ed.), Oxford Univ. Press, London, 1963.
17. J. Marsden, *Bull. Amer. Math. Soc.* **80** (1974), 154.
18. J. Marsden, "Applications of Global Analysis to Mathematical Physics," Publish or Perish, Boston, 1974.
19. J. C. S. Meng, The physics of vortex ring evolution, manuscript (1976).
20. R. Rogallo, Personal communication, 1975.
21. H. Schlichting, "Boundary Layer Theory," McGraw–Hill, New York, 1960.
22. K. Stewartson, *Quart. J. Mech. Appl. Math.* **11** (1958), 399.
23. A. Shestakov, Ph.D. Thesis, Department of Mathematics, University of California at Berkeley, 1975.
24. A. Shestakov, *in* "Proceedings of the 5th International Conference on Numerical Methods in Fluid Dynamics," Springer, New York, 1977.
25. W. W. Willmarth, *Advances in Appl. Mech.* **15** (1975), 1.

Flame Advection and Propagation Algorithms*

ALEXANDRE JOEL CHORIN

Department of Mathematics and Lawrence Berkeley Laboratory, University of California, Berkeley, California 94720

Received March 16, 1979; revised April 26, 1979

We present a simple algorithm for approximating the motion of a thin flame front of arbitrary shape and variable connectivity, which is advected by a fluid and which moves with respect to the fluid in the direction of its own normal. As an application, we examine the wrinkling of a flame front by a periodic array of vortex structures.

OUTLINE OF GOAL AND METHOD

Consider a fluid occupying a domain D with boundary ∂D, in two- or three-dimensional space. The fluid in a subdomain $D_1 \subset D$ is burned, the fluid in $D_2 = D - D_1$ is unburned, and the boundary ∂D_1 between D_1 and D_2 is transported by the velocity of the fluid and also moves with a velocity U in the direction of its own normal; D_1 is expanding while D_2 is contracting. U is the flame speed, and may depend on such parameters as the temperature of the fluid, its chemical composition, or the distance from a solid wall. D_1 and D_2 are not assumed to be connected or simply connected. The need to represent the motion of the interface between D_1 and D_2 arises in a number of combustion problems; for example, in a number of applications one can consider a flame front as a discontinuity which acts as a source of specific volume, and the induced velocity field can be computed if the location of the flame can be found accurately.

By analogy with shock dynamics, one may attempt either to follow flames explicitly as hydrodynamic discontinuities, or one may hope to have a hydrodynamical calculation locate the flames by solving the appropriate equations without any explicit allowance for the presence of a flame. The former course runs into difficulty because normals are difficult to find in a manner which is both stable and accurate, and because programming can be overwhelmingly complex in situations where flames form pockets, reconnect, etc. The latter course runs into difficulty because flame velocity, unlike shock velocity, is not determined by the basic conservation laws (see, e.g., [6, 7]) and its determination as an intrinsic part of a general program requires an accurate and expensive evaluation of chemical reaction and heat transfer rates.

In the present paper we present an alternative to both of these courses, through the

* Partially supported by the Engineering, Mathematical, and Geosciences Division of the U.S. Department of Energy under Contract W-7405-ENG-48, and by the Office of Naval Research under Contract N00014-76-0-0316.

Reprinted with permission from *Journal of Computational Physics*, Vol. 35, No. 1, 1980.

use of a Huyghens principle. For the sake of simplicity, we consider a situation in which U is a constant throughout the fluid. (The case of variable U is not essentially different.) Let D_1 be the expanding region containing burned gas. Let $\mathbf{u}_1, \mathbf{u}_2, \ldots, \mathbf{u}_n$ be a collection of vectors, with magnitudes $|\mathbf{u}_i| = U$, $i = 1, \ldots, n$, and whose directions are equidistributed on the unit sphere (or the unit circle in the case of plane flow). Consider the regions $D_1^{(1)}, D_1^{(2)}, \ldots, D_1^{(n)}$ obtained from D_1 by rigid translations with translation vectors respectively $\mathbf{u}_1 k, \mathbf{u}_2 k, \ldots, \mathbf{u}_n k$, where k is a time step. The union of the $D_1^{(l)}$, $\bigcup_{l=0}^{n} D_1^{(l)}$, $(D_1^{(0)} = D_1)$ approximates, for n large enough, the body obtained from D_1 by having the boundary of D_1 move with velocity U in the direction of its normal during the time interval k. This construction is an implementation of the classical Huyghens principle: If one takes points on the boundary of D_1, starts spherical flames expanding with velocity U from each one of the points, and then constructs the union of the volume D_1 and the volumes covered by these spherical flames, the resulting body is identical to $\bigcup D_1^{(l)}$.

The construction above requires an algorithm for performing rigid body translations and can in fact be based on any such algorithm. In the applications we have carried out. we found it convenient to use a translation algorithm based on the simple line interface advection algorithm (Noh and Woodward [14]). We shall explain this algorithm in the next section. In the following section we shall use this algorithm to implement the Huyghens principle and demonstrate that the accuracy of the resulting propagation algorithm is higher than that of the underlying advection algorithm. In a final section, we shall apply a combined advection/propagation algorithm to the analysis of the effect of intermittency on the velocity of a wrinkled thin flame in a model flow.

A Simple Line Advection Algorithm

Consider a grid with mesh length h superposed on a domain D. For simplicity, we assume D is two dimensional. The centers of the mesh cells are located at $x = ih$, $y = jh$, i, j integers (Fig. 1). A velocity field is given on the associated staggered grid (Harlow and Welch [11]); the horizontal velocity $u_{i+1/2,j}$ is given at the centers $([i+\frac{1}{2}]h, jh)$ of the vertical sides of the cells, and the vertical velocity $v_{i,j+1/2}$ is given at the centers $(ih, [j+\frac{1}{2}]h)$ of the horizontal sides. Each cell in the grid may contain burned as well as unburned fluid, and the volume fraction f_{ij} of burned fluid is given in each cell; $0 \leqslant f_{ij} \leqslant 1$. To clarify the discussion, we shall call burned fluid "black" and unburned fluid "white". The task at hand is to transport the black fluid through D with the given velocity field $\mathbf{u} = (u, v)$. This can be done only if the interface between black and white volumes can be reconstructed from the given partial volumes f_{ij}.

The ideas in the simple line interface algorithm (Noh and Woodward [14]) are as follows: An interface is drawn in each cell on the basis of an inspection of the partial volumes f_{ij} in the cell itself and in its immediate neighbors; the interface consists of horizontal and vertical lines and is made as simple as possible. The velocity at the interface is then produced from the given velocities by interpolation (in our program,

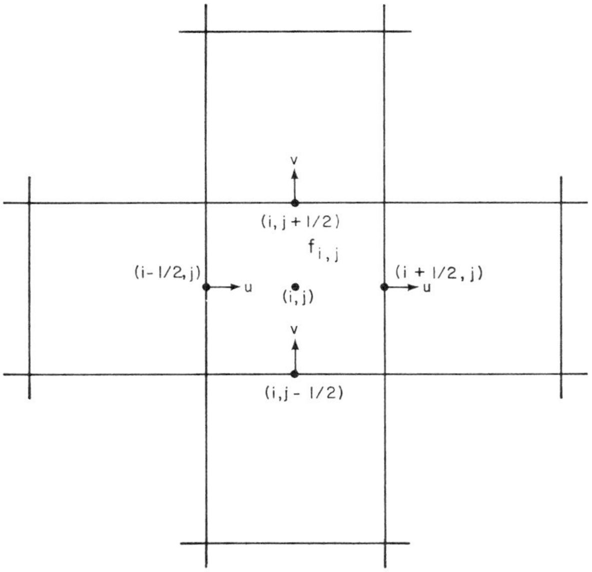

FIG. 1. Computational grid for advection.

by linear interpolation). The black volume is transported in two fractional steps, one vertical and one horizontal; the geometry of the interfaces is adapted to the direction of the flow, and it is not required that the interface constructed for the horizontal half-step coincide with the interface constructed for the vertical half-step.

Consider the horizontal half-step, and consider a cell centered at (ih, jh) with partial volume f_{ij}. We distinguish the following cases:

I. No interface. $f_{ij} = 0$ or $f_{ij} = 1$. This is the simplest and usually by far the most frequent case. The fluid in the cell moves as a whole, with the right side moving with velocity $u_{i+1/2,j}$ and the left side with velocity $u_{i-1/2,j}$. With appropriate programming, usually nothing is actually computed in this case.

II. Vertical Interface. $0 < f_{ij} < 1, f_{i+1,j} = 0$ and either $f_{i,j+1} = 0,\ f_{i,j-1} = 0$ or $f_{i,j+1} > 0, f_{i,j-1} > 0$. It is reasonable to guess that the interface is vertical and located at $x = (i - \frac{1}{2})\, h + f_{ij} h$ (Fig. 2a). The following three cases are identical, except for an interchange of the roles of right and left and/or the roles of black and white:

a. $0 < f_{ij} < 1, f_{i-1,j} = 0, f_{i+1,j} > 0$, with either $f_{i,j+1} = f_{i,j-1} = 0$ or $f_{i,j+1} > 0$, $f_{i,j-1} > 0$;

b. $0 < f_{ij} < 1, f_{i-1,j} < 1, f_{i+1,j} = 1$, with either $f_{i,j+1} = f_{i,j-1} = 1$ or $f_{i,j+1} < 1$, $f_{i,j-1} < 1$;

c. $0 < f_{ij} < 1, f_{i-1,j} = 1, f_{i+1,j} < 1$, with either $f_{i,j+1} = f_{i,j-1} = 1$ or $f_{i,j+1} < 1$, $f_{i,j-1} < 1$.

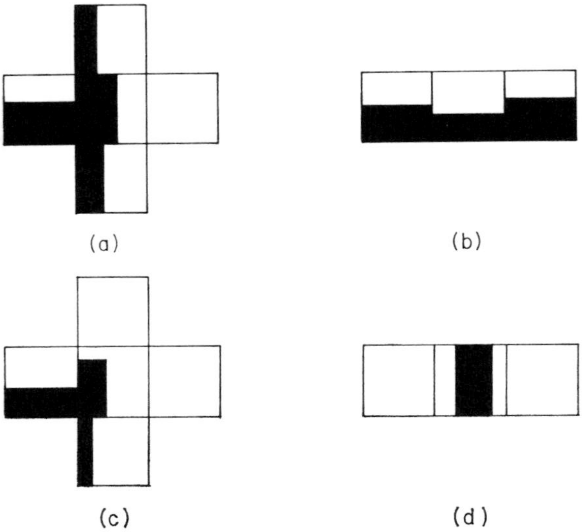

FIG. 2. Cases considered in advection algorithm.

III. Horizontal Interface. $0 < f_{ij} < 1, 0 < f_{i+1,j} < 1, 0 < f_{i-1,j} < 1$. The cell is assumed to contain a horizontal interface located at $y = (j - \frac{1}{2})\, h + f_{ij}h$ (Fig. 2b).

IV. Corner. $0 < f_{ij} < 1, 0 < f_{i-1} < 1, f_{i+1,j} = 0, f_{i,j+1} = 0, f_{i,j-1} > 0$ (Fig. 2c). The black fluid is assumed to lie in a rectangle in the lower left corner of the cell; the horizontal side of the rectangle has length a, and the vertical side has length b. We must have

$$ab = f_{i,j}h_2 .$$

We also require

$$\frac{b}{a} = \frac{f_{i-1,j}}{f_{i,j-1}},$$

whenever this equation leads to $b \leqslant h, a \leqslant h$. If this equation leads to $b > h$, we set $b = h$ and $a = f_{ij}h$; if this equation leads to $a > h$ we set $a = h$ and $b = f_{ij}h$. There are seven related cases, three of which yield black rectangles in one of the other three corners, and each of the remaining four leads to a white rectangle in one of the corners. These are obtained by appropriate interchanges of the roles of top and bottom, right and left, and black and white.

V. Thin Finger. $0 < f_{ij} < 1, f_{i+1,j} = f_{i-1,j} = 0$. The black fluid is assumed to occupy a thin finger inside the cell (Fig. 2d). The exact location of the finger is chosen at random as follows: The black finger occupies the region $a \leqslant x \leqslant b$, $a = (i - \frac{1}{2})\, h + \frac{1}{2}(1 - f_{ij})\, \theta$, $b = a + f_{ij}$, where θ is a member of a sequence equidistributed on [0, 1]. Examples of suitable equidistributed sequences can be found in Lax

[13], Chorin [4], Colella [8]. At each time half step, a new θ is chosen, but for a fixed time, the same θ is used in all cells in which this case occurs. A related case is found by exchanging the role of black and white.

The constructions in cases I, II, III were used in Noh and Woodward [14]. Their work contains additional features designed to describe effectively the motion of a fluid system with many components. Case IV is introduced here to improve the resolution of the method. Case V is important because in our application it occurs often. In [14], the finger is placed in the middle of the cell, and as a result the displacement of the finger is determined by the Courant number uk/h rather than by the velocity u (this remark is due to C. Fenimore [9]). The remedy proposed here is based on the Glimm construction [4, 10], and it ensures that on the average the motion of the finger is computed correctly. Fenimore [9] has proposed a more accurate remedy. It is known from experience with other random choice methods that the numbers θ_1 and θ_2 used in the horizontal and vertical half-steps must be independent. In the calculations to be described, we follow Colella and use two independent van der Corput sequences for the θ's.

The algorithm is stable whenever the Courant condition $(\max|u|)\,k/h < \frac{1}{2}$ is satisfied.

As an example, consider a rectangle of black fluid occupying 21 cells, transported by a "fluid" undergoing rigid body rotation centered at 0. The distance of the center of the rectangle from 0 is five cells (the problem can be scaled independently of h). In Fig. 3 we display on the right the original configuration of the black fluid, and on the

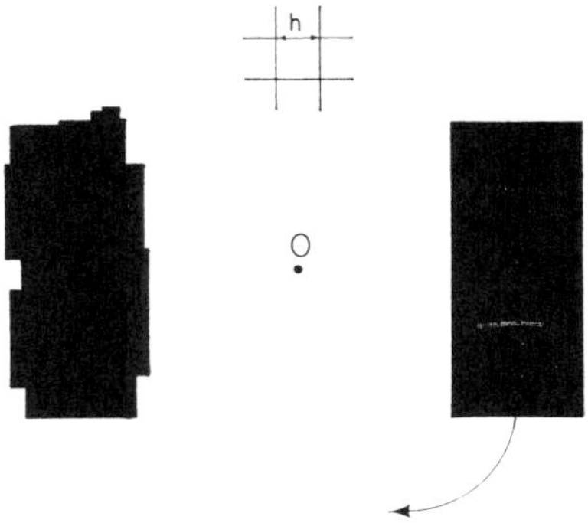

FIG. 3. An example of advection.

left the computed configuration obtained after a rotation of 180°. The lines are drawn as they are interpreted by the program. The uncertainty in the position of an interface is always less than one mesh length and, as can be expected, is largest at the corners. The accuracy is competitive with that of other methods for performing advection calculations.

Implementation of the Huyghens Principle

Consider a region $D_1 \subset D$ in the plane whose boundary is propagating with velocity U. At time $t = nk$, n integer, D_1 is described by an array of partial volumes, f_{ij}^n. Consider the 8 angles $\alpha_l = (l-1)\pi/4$, $l = 1,..., 8$, and the corresponding translation vectors $\mathbf{u}_l = (U \cos \alpha_l, U \sin \alpha_l)$. Use the algorithm described in the previous section to translate the area (described by the f_{ij}) successively by each one of the velocity fields $\mathbf{u}_l$; this results in 8 new areas $f_{ij}^{(l)}$, $l = 1,.., 8$. Write $f_{ij}^{(0)} = f_{ij}^n$, and then write

$$f_{ij}^{n+1} = \max_{0 \leqslant l \leqslant 8} f_{ij}^{(l)}.$$

This is our implementation of the Huyghens principle.

Note the following facts:

(i) Each cell in the grid has 8 neighbors. The amount of mass transported from any one cell to any one of its neighbors is largest when the translation vector points from the center of the given cell to the center of the neighboring cell. All such directions coincide with one of the directions determined by the α_l. Any additional directions are redundant and will not affect f_{ij}^{n+1}.

(ii) In three dimensional space, 26 directions are needed. The amount of resulting labor is still modest if care is taken to ensure that the calculations are performed only when they are needed, i.e., when $0 < f_{ij} < 1$ in a cell under consideration or in a neighboring cell.

(iii) In the plane a single pair of θ's in case V is sufficient for all translations during a given time step; a single triplet is needed in three dimensions.

(iv) Alternate strategies for implementing the Huyghens principle, in which fewer directions are used in conjunction with a sampling strategy for the angles, have been tried, but resulted in modest savings in computing effort with a non-negligible loss in accuracy.

The accuracy of the propagation algorithm just described was consistently higher than that of the underlying advection algorithm in all cases we ran. There are two explanations: (i) the advection algorithms are most accurate when the velocity field is one dimensional, which is the case in each one of the translations used to implement the Huyghens principle, and (ii) if the propagation algorithm underestimates or overestimates the length of the interface, the error is self-correcting to a substantial extent. As an example, we ignited the fluid in one fluid cell and followed the resulting flame propagation; in Fig. 4 we display the flame front obtained with $U = 0.2$, at $t = 1.83 = 70k$, $h = 1/19$, $Uk/h = .099$. The fractional volumes at the edge of the flames are drawn as they are interpreted by the program. The slight asymmetry reflects the effect of the θ's. The middle square is the square ignited at $t = 0$.

The original area of burned fluid is h^2, which equals the area of a circle of radius $r_0 = h/(\pi)^{1/2}$. The area of burned gas should be approximately $A = \pi(r_0 + Ut)^2$. Let A_c be the area of burned gas as computed by the program, $A_c = \sum f_{ij}h^2$. In table I we

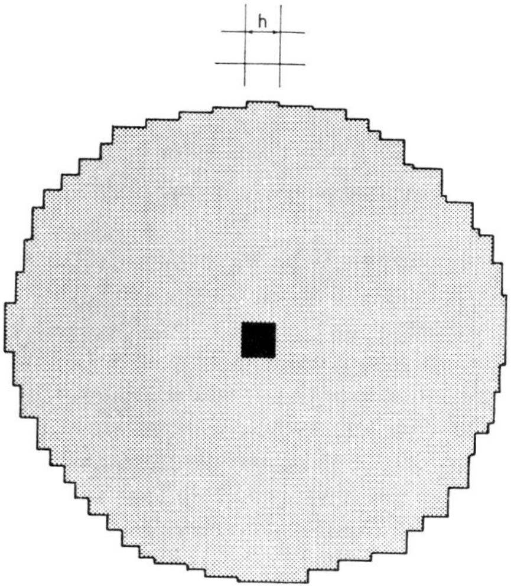

FIG. 4. Propagation of a circular flame.

TABLE I

Error in a Circular Flame Calculation, $h = 1/19$

n	A_c	$A_c - A$	$(A_c - A)/A_c$
1	0.0038	−0.000039	−0.010
2	0.0049	−0.000093	−0.018
5	0.0091	0.00064	0.065
10	0.019	0.0011	0.054
20	0.056	0.00080	0.014
30	0.110	−0.00028	−0.0025
40	0.181	−0.0013	−0.0070
50	0.269	−0.0028	−0.0085
60	0.386	−0.0033	−0.0086
70	0.497	−0.0038	−0.0077
80	0.636	−0.0044	−0.0069

display the area A_c, the error $A - A_c$, and the relative error $(A - A_c)/A_c$, with the parameters h, U, Uk/h, as above. Note that for small $t = nk$, substantial contributions to the value of $A - A_c$ are due to the fact that our formula for A is not exact, as well as to the statistical fluctuations in A_c due to the reliance on the θ's. The algorithm does perform well.

If the flame is advected by a fluid while it is propagating, the advection algorithm and the propagation algorithm can be used as successive fractional steps in the determination of the location of the front. The propagation algorithm is stable whenever the underlying advection algorithm is stable.

The Effect of Intermittency on the Velocity of Wrinkled Flames

We now present an application of the method above to a simplified problem in flame theory. (An excellent account of the subject can be found in Williams [17].) Under conditions which are often encountered in practice, one believes that a turbulent flame propagates faster than a laminar flame mainly because a turbulent velocity field wrinkles the flame and increases the area available for burning. Let the velocity of the turbulent flame be denoted by u_a, and let $u_l \equiv U$ be the velocity of an unwrinkled flame in a fluid of the same temperature and composition. It has been observed from experiments (Andrews *et al.* [1]) that in many situations the ratio u_a/u_l is roughly proportional to the intrinsic Reynolds number $R_\lambda = u'\lambda/\nu$, where u' is the rms intensity of the turbulence, λ is the Taylor microscale (for a definition, see e.g., [1, 15]), and ν is the viscosity. According to recent theories, turbulence can be usefully described as a random array of vortices [see, e.g., [4]). A theory described in [3] and experiments described in [12] lead one to believe that these vortices are rod-like, and thus a two-dimensional calculation, performed in a plane normal to the axes of these vortices, should describe their main effects. A calculation presented by Tennekes [15] suggests that λ is the order of magnitude of the diameter of these vortices.

Thus, in order to provide the simplest possible explanation of the observation of Andrews *et al.*, we are led to the following problem: Consider a time-independent periodic array of vortical structures in the plane. At $t = 0$ a plane flame front coincides with the y axis. We wish to follow the wrinkling of the flame front and the consequent increase in the velocity of the flame.

The velocity field is periodic with period $L = 1$ in both x and y directions. Consider one periodic box, $-\frac{1}{2} \leqslant x \leqslant \frac{1}{2}$, $-\frac{1}{2} \leqslant y \leqslant \frac{1}{2}$. Consider the velocity field given by $\mathbf{u} = (u, v)$, $u = -\partial_y \psi$, $v = \partial_x \psi$, where $\psi = C \exp(-(x^2 + y^2)/\lambda^2)$. λ is the "microscale." $\mathbf{u}$ is not periodic, and although it does satisfy the equation div $\mathbf{u} = 0$, it does not satisfy the discrete equations $D\mathbf{u} \equiv u_{i+1/2,j} - u_{i-1/2,j} + v_{i,j+1/2} - v_{i,j-1/2} = 0$. $D\mathbf{u} = 0$ guarantees that the area occupied by burned gas increases only due to burning (except for possible small errors due to the interpolations used in the advection algorithm). The component of $\mathbf{u}$ which is periodic and satisfies the equations $D\mathbf{u} = 0$ is obtained by the projection algorithm described in [2]. The constant C is then adjusted so that

$$u' = \left(\sum_{\text{box}} (u^2 + v^2)\, h^2 \right)^{1/2} = 1.$$

In Fig. 5 we display a typical flow configuration in a periodic box. At $t = 0$, the flame coincided with the left wall of the box. The front is shown at $t = 1.53$, $n = 125$,

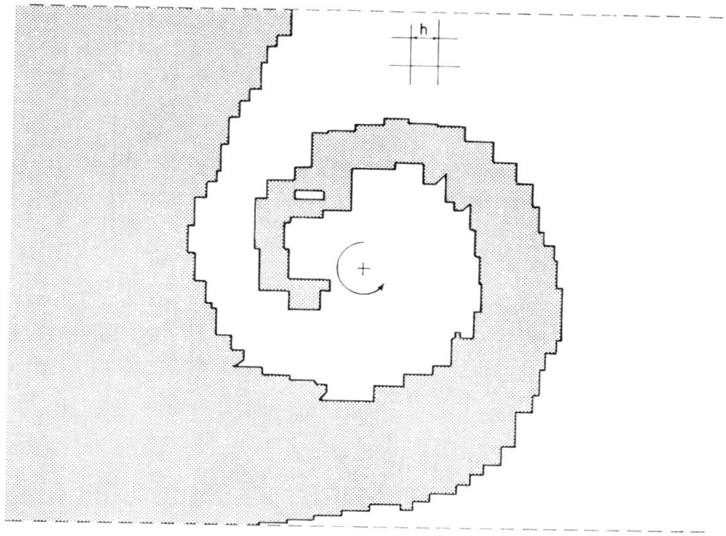

FIG. 5. Stretching of a flame by a vortical structure.

with $u_l = 0.2$, $\lambda = 0.2$, $h = 1/19$. The partial volumes are drawn as they are interpreted in a horizontal sweep, and occasional ambiguities are removed by diagonal lines.

The viscosity ν does not appear explicitly in our model; indeed, ν governs the rate at which vortical structures appear and disappear, and in our problem they do neither.

Let $A(t)$ be the portion of the periodic box occupied by burned fluid. Define $u_a = dA/dt$. Simple scaling arguments show that u_a/u_l can be a function of the ratios $u'/u_l = 1/u_l$ and $\lambda/L = \lambda$ only. Thus the analogue of the law of Andrews *et al.* is

$$\frac{u_a}{u_l} = \text{constant} \times \frac{u'}{u_l}\frac{\lambda}{L} = \text{constant} \times \frac{\lambda}{u_l}.$$

However, the original law $u_a/u_l \sim u'\lambda/\nu$ and the new law $u_a/u_l \sim \lambda/u_l$ are essentially different, since the latter implies that $u_a \sim \lambda$ independently of u_l. This last conclusion is untenable and disappears only if it can be shown that u_a/u_l is roughly independent of u'/u_l. For $u' \gg u_l$, this last statement is indeed true. In Fig. 6 we display u_a/u_l as a function of the appropriately scaled time $t^* = tu_l/0.2$ for several values of u_l. The curves coincides to a large extent, showing that u_l does not affect greatly the generation of new surface by vortical motion.

It is clear that if u_a/u_l is roughly independent of u'/u_l, the generation of new surface is roughly proportional to the scale λ of the vortical structures. In Fig. 7 we display the variation of u_a/u_l with λ as a function of time. It can be seen that for a given t the value of u_a/u_l is indeed roughly linear in λ. u_a/u_l increases when the vortex

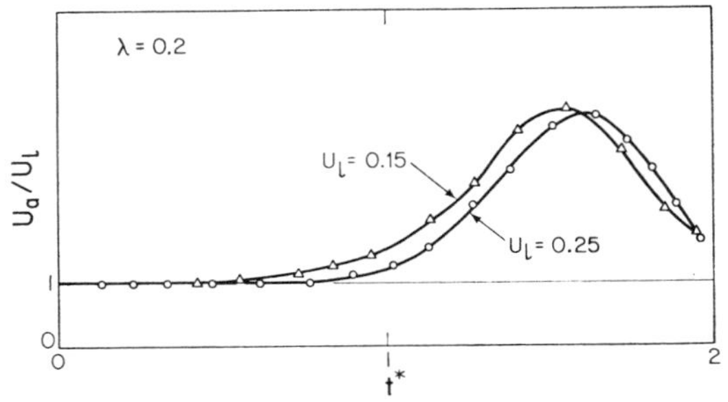

FIG. 6. Effect of u_l on u_a/u_l.

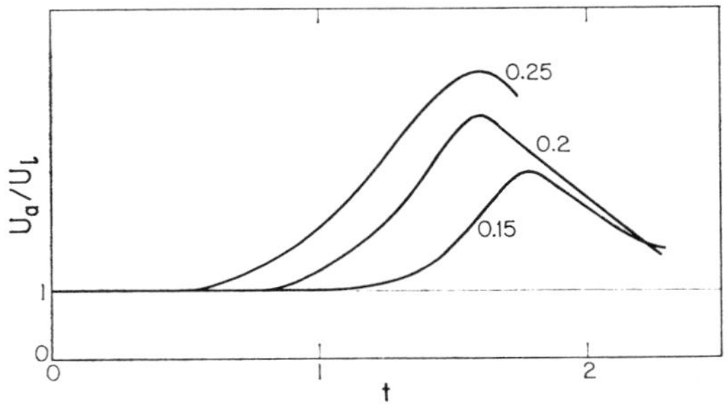

FIG. 7. Effect of scale on u_a/u_l.

meets the flame, then decreases when the flame consumes the newly added flame length. The calculation was stopped when the flame was overflowing the box. Thus, in the narrow confines of our model problem, we have a reasonable explanation of the observation offered in [1].

ACKNOWLEDGMENT

I would like to thank Mr. William Noh for helpful discussions and for introducing me to the simple line algorithm.

REFERENCES

1. A. E. ANDREWS, D. BRADLEY, AND S. LWAKABAMBA, *Combust. Flame* **24** (1975), 285.
2. A. J. CHORIN, *Math. Comp.* **22** (1968), 745.
3. A. J. CHORIN, *in* "Proceedings, Second International Conference on Numerical Methods in Fluid Dynamics" (M. Holt, Ed.), Springer-Verlag, 1970.
4. A. J. CHORIN, *J. Fluid Mech.* **63** (1974), 21.
5. A. J. CHORIN, *J. Comp. Phys.* **25** (1977), 253.

6. A. J. Chorin and J. E. Marsden, "A Mathematical Introduction to Fluid Mechanics," Springer-Verlag, New York, 1979.
7. R. Courant and K. O. Friedrichs, "Supersonic Flow and Shock Waves," Interscience, New York, 1948.
8. P. Colella, Ph.D. thesis, University of California, Berkeley, Mathematics Department, 1979.
9. C. Fenimore, private communication, 1978.
10. J. Glimm, *Comm. Pure Appl. Math.* **18** (1965), 697.
11. F. H. Harlow and J. E. Welch, *Phys. Fluids* **8** (1965), 2182.
12. A. Y. S. Kuo and S. Corrsin, *J. Fluid Mech.* **56** (1972), 447.
13. P. D. Lax, *SIAM Rev.* **11** (1969), 7.
14. W. Noh and P. Woodward, *in* "Proceedings, Fifth International Conference on Fluid Dynamics" (A. I. van de Vooren and P. J. Zandbergen, Eds.), Springer-Verlag, 1976.
15. H. Tennekes, *Phys. Fluids* **11** (1968), 669.
16. H. Tennekes and J. L. Lumley, "A First Course in Turbulence," MIT Press, Cambridge, Mass./London, 1972.
17. F. A. Williams, "Combustion Theory," Addison–Wesley, Reading, Mass., 1965.

VORTEX MODELS AND BOUNDARY LAYER INSTABILITY*

ALEXANDRE JOEL CHORIN†

Abstract. Random vortex methods are applied to the analysis of boundary layer instability in two and three space dimensions. A thorough discussion of boundary conditions is given. In two dimensions, the results are in good agreement with known facts. In three dimensions, a new version of the method is introduced, in which the computational elements are vortex segments. The numerical results afford new insight into the effects of the third dimension on the stability of a boundary layer over a flat plate.

Key words. *vortex, boundary layers, random walk, three-dimensional instability*

Introduction. The random vortex method as described in Chorin [7] is intended for the approximation of flows at high Reynolds number R. Its main features are as follows: (i) the nonlinear terms in the Navier–Stokes equation are taken into account by a detailed analysis of the inviscid interactions between vortices of small but finite core ("vortex blobs"), (ii) viscous diffusion is taken into account by adding to the motion of the vortices a small random Gaussian component of appropriate variance, and (iii) no-slip boundary conditions are approximated by a vorticity creation algorithm. Fuller details are given below. Developments, modifications, and applications of the method can be found e.g. in Ashurst [1], Chorin [10], [11], Leonard [26], [27], McCracken and Peskin [30], Shestakov [36]. Theoretical analysis can be found in Hald [18], Hald and Del Prete [19], and Chorin et al. [12].

This grid-free method is suitable for the analysis of flow at high Reynolds number because it has no obvious intrinsic source of diffusion. Most approximation methods solve equations which are close to the equations one wants to solve; the difference consists of higher order terms multiplied by small parameters. This is also the form of the diffusion term, and as a result, in most methods, the effects of a small R^{-1} are dominated by numerical effects and the physics of high Reynolds number flow are suppressed. In vortex methods, the misrepresentation of the higher harmonics which occurs in the usual discretization methods (which usually has a diffusive effect among other effects) is replaced by the misrepresentation of the interaction of neighboring vortices (an essentially inviscid phenomenon which is a source of error, but not of diffusive error). In the absence of the nonlinear terms, the diffusion is approximated on the average exactly. Thus one may hope that the results of the calculation approximate the flow at whatever Reynolds number was intended, albeit with a statistical error, rather than at some other lower Reynolds number intrinsic to the algorithm. A good guess at the solution of the problem one wants to solve is better than an unambiguous solution of the wrong problem.

The method produces a flow field which is random. The error in the calculation is the sum of two parts: the expected value of the computed solution differs from the true solution, and any realization of the computed solution (or more accurately, any functional thereof) differs from the expected value by a random amount which can be estimated by its standard deviation (see e.g., Lamperti, [25]). The expressions for these quantities will be given below, when the appropriate notation will be available.

In the present paper we apply random vortex methods to the analysis of the boundary layer over a flat plate in two and three space dimensions. The calculations

* Received by the editors May 15, 1979. This work was supported in part by the Engineering, Mathematical, and Geosciences Division of the U.S. Department of Energy through the Lawrence Berkeley Laboratory.

† Department of Mathematics, University of California, Berkeley, California 94720.

Reprinted with permission from *SIAM Journal on Scientific and Statistical Computing*, Vol. 1, No. 1, 1980.

have two main objects. In the two dimensional case we shall show that the vortex method exhibits a physical instability at an appropriate Reynolds number. The ability to do so is of course a basic requirement for any method which claims to have some use at high Reynolds number. The specific problem we apply our method to has a simplifying feature, inasmuch as the location of the sharp gradients is known in advance to be near the wall, and thus the equations of motion can be solved in two dimensions by finite difference or other non-statistical methods in appropriately scaled variables. The interesting fact about our calculation is that it does not require such preliminary scaling of the variables, i.e., the random walk can be relied upon to create the appropriate diffusive length scale.

The second main goal of our calculation is to use the method to investigate the much harder problem of boundary layer instability in three dimensions, and in particular, two of the striking features of its solution: The formation of streamwise vortices and the creation of active spots. The three dimensional calculation requires a generalization of our method, and both the two dimensional and three dimensional problems afford the opportunity to use an improved algorithm for imposing the boundary conditions accurately.

In the next four sections we present the calculation in two dimensions. In later sections we present the three dimensional calculations.

The physical problem in two dimensions. Consider a semi-infinite flat plate placed on the positive half-axis, with an incompressible fluid of density 1 occupying the half space $y \geqq 0$. At time $t<0$ the fluid is at rest. At $t=0$, the fluid is impulsively set into motion with velocity U_∞. The flow is described by the Navier–Stokes equations,

$$\partial_t \xi + (\mathbf{u}\cdot\nabla)\xi = R^{-1}\Delta\xi, \tag{1a}$$

$$\Delta\psi = -\xi, \tag{1b}$$

$$u = \partial_y \psi, \qquad v = -\partial_x \psi, \tag{1c}$$

where $\mathbf{u} = (u, v)$ is the velocity vector, $\mathbf{r} = (x, y)$ is the position vector, ξ is the vorticity, ψ is the stream function, $\Delta \equiv \nabla^2$ is the Laplace operator, and R is the Reynolds number, $R = U_\infty L/\nu$, where L is a length scale typical of the flow. The boundary conditions are

$$\mathbf{u} = (U_\infty, 0) \quad \text{at } y = \infty,\ t>0, \tag{1d}$$

$$u = v = 0 \quad \text{for } y = 0,\ x>0, \tag{1e}$$

$$\frac{\partial v}{\partial y} = 0 \quad \text{for } y = 0,\ x<0. \tag{1f}$$

Initially, $\mathbf{u} = (U_\infty, 0)$ everywhere.

If R is large, the Prandtl boundary layer equations should provide a reasonable description of the flow near the plate and away from the leading edge. These equations can be written in the form [Schlichting [35], Chorin [10], [11]],

$$\partial_t \xi + (\mathbf{u}\cdot\nabla)\xi = \nu \partial_y^2 \xi, \tag{2a}$$

$$\xi = -\partial_y u, \tag{2b}$$

$$\partial_x u + \partial_y v = 0. \tag{2c}$$

where ξ, u, v, x, y have the same meaning as in equations (1), and ν is the viscosity. If $U_\infty = 1$ and $L = 1$, $R = \nu^{-1}$. The boundary conditions for equations (2) are: $u = U_\infty$ for $y = \infty$, $\mathbf{u} = 0$ for $y = 0$. Equations (2) have a stationary solution, the Blasius solution, which is a function of the similarity variable $\mu = y/\sqrt{x\nu}$. Let the displacement thickness δ

be defined by

$$\delta = \int_0^\infty (1 - u/U_\infty)dy;$$

the corresponding Reynolds number is $R_\delta = U_\infty \delta/\nu$. In Blasius flow, $\delta = 1.72\sqrt{\nu x}$, and $R_\delta = 1.72\sqrt{x/\nu}$, where it is assumed that $U_\infty = 1$. δ and R_δ are increasing functions of x. For $R_\delta \geqq R_{\delta c}$ the Blasius solution is unstable to infinitesimal perturbations which satisfy equations (1) (see Lin [29]); $R_{\delta c} \simeq 520$, (See Jordinson [21]). These unstable modes are the Tollmien–Schlichting waves. The vortex interpretation of the waves is as follows: The boundary layer is a region of distributed vorticity imbedded in a shear flow. Vorticity imbedded in a shear tends to become organized into coherent macroscopic structures ("negative temperature states", "local equilibria", see Onsager [32], Chorin [8]). This tendency is counteracted by the diffusive effects. The latter become weaker as x increases, since the vorticity gradients decrease as the layer spreads. Far enough downstream (i.e., for R_δ large enough), the tendency towards coherence can overcome the diffusive effects; the Tollmien–Schlichting waves can be viewed as a weak train of organized vortex structures.

The value of $R_{\delta c}$ given above has to be lowered if the unperturbed flow is treated as a nonparallel flow and if edge effects are taken into account (Townsend [37]). More importantly, the boundary layer is unstable to perturbations of a finite amplitude for values of R_δ smaller than $R_{\delta c}$ (for analysis of similar situations, see Eckhaus [14], Meksyn and Stuart [31]). A survey of finite amplitude stability theory for the flat plate problem is given by Roshotko [33]). The boundary layer becomes more unstable if the outside flow is turbulent or contains vortical structures (see Schlichting [35], Rogler and Reshotko [33]). Since our calculation will by its very nature contain finite amplitude perturbations, vortices, a substantial amount of noise, and edge effects, the appropriate value of R_δ which separates stable from unstable regimes is unclear. Presumably, there exists a value $R'_{\delta c}$ such that for $R_\delta \leqq R'_{\delta c}$ all perturbations decay; the best guess of $R'_{\delta c}$ we can obtain by looking at the references above is $R'_{\delta c} \simeq 300$, with a substantial margin of error. Cebeci and Smith [4] suggest a value $R'_{\delta c} \simeq 320$.

For $R_\delta \geqq R_{\delta c}$, the perturbations can grow, but I found little information as to what they do in two dimensions; presumably they grow and reach some finite amplitude equilibrium; this is the typical situation in other two-dimensional stability problems, for example in the thermal convection problem (see e.g. Chorin [6]). All experimental studies I know deal with the more important and more realistic three dimensional problem which will be discussed further below.

The numerical methods in two dimensions. Consider first the Navier–Stokes equations (1) in the whole plane. Assume that $\xi = \sum_j \xi_j$, where the ξ_i are functions of small support (ξ_i is a "blob"). Let $\psi = \sum \psi_j$, where $\Delta\psi_j = -\xi_j$. (If we had $\xi_j = \kappa_j \delta(\mathbf{r} - \mathbf{r}_j)$, $\kappa_j =$ constant, we would have concluded that $\psi_j = -(\kappa_j/2\pi)\log|\mathbf{r} - \mathbf{r}_j|$.) For ξ_j smooth but of small support, let $\kappa_j \equiv \int \xi_j \, dxdy$, and we must have

$$\lim_{|\mathbf{r}| \to \infty} \frac{\psi_j}{(1/2\pi)\log|\mathbf{r} - \mathbf{r}_j|} = -\kappa_j.$$

For $|\mathbf{r} - \mathbf{r}_j|$ small, ψ_j differs from $(\kappa_j/2\pi)\log|\mathbf{r} - \mathbf{r}_j|$ (or else it would introduce undesirable unbounded velocities, see Chorin [7], Hald [18]). We set

$$\psi_j(\mathbf{r}) = \begin{cases} \dfrac{\kappa_j}{2\pi}\log|\mathbf{r} - \mathbf{r}_j|, & |\mathbf{r}| \geqq \sigma, \qquad (3a) \\[2ex] \dfrac{\kappa_j}{2\pi}\dfrac{|\mathbf{r}|}{\sigma} + \text{const.}, & |\mathbf{r}| < \sigma. \qquad (3b) \end{cases}$$

This is the form introduced in Chorin [7]; it differs from the forms described by Hald in [18] for reasons which will become apparent below. Clearly $\xi_j = -\Delta\psi_j$ is of small support. σ is a cut-off which remains to be determined.

Equations (1) state that the vorticity moves with the velocity field which it induces, i.e., let $\mathbf{u}_j = (u_j, v_j)$ be the velocity field induced by the jth blob, and let $\mathbf{r}_i = (x_i, y_i)$ be the center of the ith blob, then

$$\frac{d\mathbf{r}_i}{dt} = \sum_{j\neq i} \mathbf{u}_j, \qquad (\mathbf{u}_j \text{ evaluated at } r_i).$$

This equation can be approximated by

$$\mathbf{r}_j^{n+1} = \mathbf{r}_i^n + k \sum_{j\neq i} \mathbf{u}_j \tag{4}$$

where k is a time step and $\mathbf{r}_i^n \equiv \mathbf{r}_i(nk)$. Hald [19] has shown that a higher order method is indeed more accurate but we shall use (4) for the sake of simplicity.

The heat equation is well known to be solvable by a random walk algorithm (see Chorin [7]). As a result, equations (1) can be solved by moving the blobs according to the law

$$\mathbf{r}_i^{n+1} = \mathbf{r}_i^n + k \sum_{j\neq i} \mathbf{u}_j + \boldsymbol{\eta} \tag{5}$$

where $\boldsymbol{\eta} = (\eta_1, \eta_2)$, η_1, η_2 independent Gaussian random variables with mean 0 and variance $2k/R$.

Suppose we wish to solve equations (1) in a domain D with boundary ∂D. The normal boundary condition $\mathbf{u} \cdot \mathbf{n} = 0$ on ∂D, $\mathbf{n}$ normal to ∂D, can be readily taken into account by solving $\Delta\psi = -\xi$ subject to the appropriate boundary condition, with the help of potential theory. In the case of flow over a flat plate, the method of images will do the job. The no-slip boundary condition $\mathbf{u} \cdot \mathbf{s} = 0$, $\mathbf{s}$ tangent to ∂D, can be imposed through the creation of the appropriate amount of vorticity: Let u_0 be the velocity component tangent to the wall created by the algorithm as described so far, and suppose $u_0 \neq 0$. The no-slip condition and the viscosity will create a boundary layer in which the total vorticity per unit length is

$$\int_{\text{wall}}^{\text{interior}} \xi \, dn = \int \frac{\partial u}{\partial n} \, dn = u_0.$$

In the algorithm presented in [7], we reproduced this effect numerically by creating a vortex sheet of strength u_0 at the wall, dividing its vorticity among blobs, and allowing these blobs to participate in the subsequent motion of the blobs according to the laws (5). If a blob is created at every piece of boundary of length h, its intensity is

$$\kappa = u_0 h. \tag{6}$$

If a blob inside the fluid happens to cross the boundary, it is removed. It should be apparent that the amount of vorticity created at each time step depends on the cut-off σ. If σ is small, the backwash of the vortex may be large, and a vortex whose center is near the boundary will create a vortex whose intensity will have an opposite sign, etc. If σ is large, the backwash of a newly created vortex may not be sufficient to annihilate u_0, and more vortices will be created, all of the same sign. Presumably, on the average the total amount of vorticity is independent of σ. The algorithm in this form is not accurate (see Chorin et al. [12]). This lack of accuracy as well as the desire to reduce the amount of computational labor have led to the formulation of the vortex sheet algorithm with which one can solve the boundary layer equations (2) (Chorin [10]). The computational elements are segments of a vortex sheet. Let u_0 be the velocity component parallel to the wall. A segment S of a vortex sheet is a segment of a straight line, of length h, parallel to the wall, such that u above S differs from u below S by an amount ξ; ("above" means "further from the wall"), $u_{\text{above}} - u_{\text{below}} = \xi$. ξ is the intensity of the sheet.

Consider a collection of N segments S_i, with intensities ξ_i, $i = 1, \cdots, N$. Let the center of S_i be $\mathbf{r}_i = (x_i, y_i)$. To describe their motion, one begins with equations (2b) and (2c). Equation (2b) can be integrated in the form

(7a) $$u(x, y) = U_\infty - \int_y^\infty \xi(x, \alpha)\, d\alpha,$$

where U_∞ is the velocity at infinity seen by the layer. Equation (2c) yields

(7b) $$v(x, y) = -\partial_x \int_0^y u(x, \alpha)\, d\alpha$$

Equations (7a) and (7b) allow one to determine u, v if $\xi = \xi(x, y)$ is given. One can visualize each sheet as casting a shadow between itself and the wall. The darker the shadow, the smaller u becomes. Whatever fluid enters a shadow region from the left and cannot leave on the right must leave upwards. From equations (7) one can derive the following expression for $\mathbf{u}_i = (u_i, v_i)$ at the center $\mathbf{r}_i$ of the ith sheet

(8a) $$u_i = U_\infty - \tfrac{1}{2}\xi_i - \sum_j \xi_j d_j,$$

where $d_j = 1 - |x_i - x_j|/h$ is a smoothing function, and the summation is over all S_j for which $0 \leqq d_j \leqq 1$ and $y_j \geqq y_i$. This is of course a small subset of all the sheets; only the sheets which lie in a narrow vertical strip around u_i affect u_i. Similarly,

(8b) $$v_i = -(I_+ - I_-)/h,$$

where

(8c) $$I_\pm = U_\infty - \textstyle\sum_\pm \xi_j d_j^\pm y_j^*,$$

(8d) $$d_j^\pm = 1 - |x_i \pm h/2 - x_j|/h,$$

(8e) $$y_j^* = \min(y_i, y_j).$$

The sum $\sum_+$ (resp. $\sum_-$) is over all S_i such that $d_j^+ \leqq 1$ (resp. $d_j^- \leqq 1$). The motion of the sheets is then given by

(9a) $$x_i^{n+1} = x_i^n + k u_i,$$

(9b) $$y_i^{n+1} = y_i^n + k v_i + \eta.$$

These formulas are analogous to (4); η is a Gaussian random variable with mean 0 and variance $2\nu k$; it appears only in the y component because equations (2) take into account diffusion in the y direction only.

This vortex sheet algorithm generates a velocity field $\mathbf{u} = (u, v)$ which satisfies the boundary condition $u = U_\infty$ at $y = \infty$, $v = 0$ at $y = 0$. The no-slip boundary condition $u = 0$ at $y = 0$ can be satisfied by the following vorticity generation procedure (see [10]): Continue the flow from $y > 0$ to $y < 0$ by antisymmetry, i.e., $\mathbf{u}(x, -y) = -\mathbf{u}(x, y)$. Since $\xi = -\partial u/\partial y$, and both u and y change signs, we have $\xi(x, -y) = \xi(x, y)$; if $u(x, 0) = u_0 \neq 0$, we also have a vortex sheet of strength $2u_0$ at the wall. This sheet can be divided into segments and allowed to participate in the subsequent motion. The antisymmetry can be imposed by reflecting any sheet which crosses the wall back into the fluid. One can require that all the sheets created satisfy the requirement $|\xi_i| \leqq \xi_{\max}$, where $\xi_{\max}$ is some reasonably small quantity. To do this, one may have to create more than one sheet at any one point at any given time step. The sheet method can be modified to make it more efficient and to reduce the variance of the results (see [10]). The interaction of the sheets is not singular and no cut-off is needed. The amount of vorticity created at the wall is unambiguous, and the cost of the calculation is small. This is of course balanced by the fact that the Prandtl equations are not uniformly valid approximations to the Navier–Stokes equations, and the transition from sheets to blobs involves in general a decision process which in turn is not unambiguous.

Note that the antisymmetry just described cannot be used directly with the vortex blob method. Indeed, if $\mathbf{u}(x, -y) = -\mathbf{u}(x, y)$, it does not follow in general that

$$\xi(x, -y) \equiv \left(-\frac{\partial u}{\partial y} + \frac{\partial v}{\partial x}\right) \quad \text{at } (x, -y) = \xi(x, y),$$

since x does not change sign. Thus, to impose the boundary conditions accurately on the blob method we shall have to use the sheet method as a transition near the wall, see below.

The version of the sheet method that we shall use is almost identical to the one described in [10] and documented in detail in Cheer [5]; this includes tagging and variance reduction techniques. The only difference is the following: In the earlier program, sheets were created at the wall, and on the average, half of them disappeared at each step. In the present program, we make exactly half of them disappear at each step and this reduces the total number of sheets retained. This is accomplished as follows: At each point at which sheets are created, their intensity is adjusted so that their number is even. A rejection technique (Handscomb and Hammersley [20]) is then used to ensure that any two successive η's used at the wall will have differing signs. This rejection technique can be used only at the wall, or else it would destroy the independence of the successive η's in the interior and thus fail to describe the diffusion process correctly.

The sheets and the blobs are objects of a very similar nature; they are determined by the same parameters, position and intensity. A computational element (x_i, y_i, ξ_i) can be treated as either a sheet or a blob, depending on the circumstances. A sheet of negative intensity casts a shadow which slows the fluid under it; by the equation of continuity, this creates an upward flow to the left and a downward flow to the right, just as if the sheet were a vortex. The circulation around a sheet of intensity ξ is ξh, and if the sheet becomes a blob, the latter's intensity must be $\kappa = \xi h$, in agreement with equation (5).

These facts can be used to create a transition between the blobs and the wall. Pick a length l such that a blob has a small probability of jumping more than $2l$ in one random jump, i.e. l a multiple of the standard deviation $\sqrt{2k/R}$ of η. Any vortex which finds itself less than l from the boundary (inside or outside) becomes a sheet and is reflected accordingly, and also taken into account accordingly when u_0 is computed. If a blob is further outside the domain than l it is removed (presumably this happens rarely). If a sheet is inside the domain and its distance from the boundary is more than l it becomes a blob again.

The cut-off σ remains to be determined. A natural condition to impose is the following: consider a collection of blobs. As they approach each other and the boundary, their interaction should converge to the interaction of the corresponding sheets. Consider a sheet of intensity ξ at (X, Y), as well as vortex of intensity ξh at (X, Y), together with its image vortex at $(X, -Y)$ required to satisfy the boundary conditions (the sheets need no images). If $\sigma = h/\pi$, the velocity fields induced along the vertical line $x = X$ are identical (Fig. 1). The lateral effects will tend to each other as $y \to 0$. Thus, if $\sigma = h/\pi$, the interaction of the blobs will approximate the interaction of the sheets when the blobs approach the boundary. Hence $\sigma = h/\pi$ is a natural choice for σ. Note that the form (3) of ψ ensures that for $|\mathbf{r}| \leqq \sigma$ the magnitude of $\mathbf{u}$ is constant. This is the reason (3) is used. Remarks: (i) the value of σ is twice the value used in [7]. (ii) The choice of σ has the greatest effect near the wall, and thus it is natural to determine the value of σ by considering what happens near the wall. (iii) Our value of σ is large compared to the mean distance between blobs which is of order $R^{-1/2}$; this is in agreement with the requirements in Hald's proof. In summary the computational elements should be viewed as sheets near the wall, and as blobs far from the wall.

FIG. 1. *Sheets and vortices near a wall.*

A heuristic error analysis in [7] provides error estimates for the expected value of the velocity field produced by our methods in the form: error $= O(k) + O(R^{-1/2})$, R = Reynolds number based on a velocity and length scales typical of the flow away from the wall. Hald's analysis of the inviscid case suggests that this could be reduced to $O(k^2) + O(R^{-1/2})$ if the time integration were carried out more accurately. The standard deviation of a smooth functional of the velocity should be $O(R^{-1/2})$.

Application of the numerical method in two dimensions. In this section we describe the application of the vortex methods to the specific problem at hand. Note that if the sheet method is used by itself on the flat plate problem and if it converges in the mean to a stationary solution of the Prandtl equations (2); that solution is a function of the similarity variable μ only; more specifically, if two computer runs are made, with the same numerical parameters k, h, $\xi_{\max}$, etc., the same sequence of random numbers, and the same impulsive initial conditions, but with two distinct values of ν, the resulting computed solutions will be identical for equal values of $y/\sqrt{x\nu}$ and x. These facts are straightforward consequences of equations (8) (see Chorin and Marsden [11]). As a consequence, the instability of the boundary layer cannot be seen with the sheet method, and our main tool will be the blob method. We shall use the sheet method for the following limited purposes: (i) to provide a rational argument in favor of the value $\sigma = h/\pi$; (ii) as a vorticity creation algorithm, (iii) as a way of imposing an approximate Blasius flow before allowing unstable modes to grow; and (iv) as a diagnostic tool.

The number of vorticity elements required to describe the flow is large, since enough of them must be included to resolve the Tollmien–Schlichting waves, and those have a short wave length. From linear stability theory (see e.g. [29], [21]) one finds that the wave number of unstable Tollmien–Schlichting waves is between roughly $0.1/\delta$ and $0.4/\delta$ for moderate values of R_δ, say very roughly $0.3/\delta = 0.2/\sqrt{x\nu}$. The corresponding wave length is $\sim 10\pi\sqrt{x\nu}$; the number of waves between 0 and x is roughly x divided by $10\pi\sqrt{x\nu}$, i.e. $\sim R_\delta/50$. The first unstable modes occur when $R_\delta \sim 500$, i.e., one has to be able to resolve at least 10 waves between the leading edge and the first occurrence of growing modes. One can also see that the time period is correspondingly small. For this reason stability calculations based on the Navier–Stokes equations are very expensive indeed (see e.g. Fasel [15]).

There is an additional constraint in the present work. It is interesting to compare the behavior of the growing modes in two dimension with the corresponding behavior in three dimensions; the two cases are quite different, and the contrast is very instructive when one is interested in the transition to turbulence. We wish to use comparable numerical parameters in two and in three dimensions, so that the comparison of the results be believable; the cost of three dimensional calculations is of course much larger even than the cost of two dimensional calculations; we must therefore look for ways of representing the boundary layer which are as economical as possible and yet exhibit a correct behavior.

There is no obvious way in which the steady Blasius profile can be imposed exactly on our array of vortex elements at the initial time. On the other hand, a calculation which starts from impulsive initial data contains a large and rather long-lived transient component whose behavior is not easily distinguishable from that of a growing mode. Part of this problem can be removed as follows: Start the calculation by using the sheet representation only (which is cheap and allows no instability), and run for a time $0 < t < T_0$, T_0 large enough so that the Blasius profile will have been reached with some not unreasonable accuracy. At time $t = T_0$ allow some or all of the sheets to become blobs. In all the two dimensional runs described below we set $T_0 = 1$.

It is quite obvious that we shall not be able to duplicate the results of linearized stability theory. The initial data will not coincide exactly with the Blasius solution. The perturbations will not be small. In Fasel [15] the perturbation amplitude was about 0.05 of the free-stream velocity—an impossibly low level for our method. Our results should be compared with the behavior of finite amplitude perturbations in noisy flow. The advantages of our numerical method can be seen from the fact that the method requires no scaling. The very same program can be used to solve an interior flow problem. The algorithm provides its own scaling and concentrates the computing effort where it is needed. This should be particularly important in other problems where thin shear layers occur at locations which are not known in advance.

In the calculations described below, the vorticity is created at walls in the form of sheets, with all $|\xi_i| \leqq \xi_{\max}$. If the amount of vorticity needed to satisfy the boundary conditions is less than ξ_0, no sheets are created; here, $\xi_0 = \xi_{\max}/2$. When sheets find themselves at $y > l$ at time $t > T_0$, they become blobs; they become sheets again if $y < l$. l must be such that the probability that $\eta > 2l$ is small. We checked that as long as $l \sim 1.5\times$ the standard deviation of η, the results were insensitive to the value of l. Detailed calculations were performed for $0 \leqq x \leqq 1$; i.e., the typical streamwise length L is 1, and thus $R = U_\infty L/\nu = \nu^{-1}$. Both sheets and blobs were followed for $x > 1$ but allowed to move only with the random component in their laws of motion. When they reached $x = X$ they were deleted. This was done to ensure that the right boundary at $x = 1$, which is introduced only for computational convenience, behaves as an absorbing boundary and does not affect adversely the calculations in the region of interest $0 \leqq x \leqq 1$. We usually picked $X = 2$.

The interaction of two elements at least one of which was a sheet, was computed as if both were sheets. Two blobs interacted as blobs. In the computation of the tangential velocity at the wall, all elements were treated as sheets.

After much experimentation we picked $\xi_{\max}=0.6$. This is a large value of $\xi_{\max}$ and produces a crude and noisy boundary layer; however, it is sufficient for exhibiting the main effects. A relatively large value of $\xi_{\max}$ reduces the number of elements in the calculation, and, as explained above, this is of particular importance since we intend to present a three dimensional calculation. The choices of h and k are described in the next section.

In the steady state, the drag $D(x)$ on the piece of boundary between 0 and x can be computed by the momentum defect formula (see e.g. [35]).

$$D(x)=\int_0^\infty u(U_\infty-u)\,dy, \qquad u=u(x,y). \tag{10a}$$

The normalized drag is defined as

$$d(x)=D(x)/D_0(x), \tag{10b}$$

where $D_0(x)$ is the Blasius drag $D_0(x)=0.6641\sqrt{x\nu}$, which can be obtained from the Blasius solution. The velocities for use in formulas such as (10) are computed as if all the elements were sheets. We shall use $d(x)$ defined by (10) as a measure of the amplitude of the growing modes even when the flow is not steady and $D(x)$ is not really the drag on $[0, x]$.

Finally, we observed that if k was too large the solution exhibited large oscillation of no possible physical significance. This is readily understood. We are solving a moderately large system of ordinary differential equations by Euler's method. The remedy is to reduce k. $k\leqq h$ is adequate.

Numerical results in two dimensions. In Table 1 and Fig. 2 we display the normalized "drag" $d(x)$ at $X=1/2$ as a function of R and t. ($d(X)$ is the ratio $D(X)/D_0$, see formula (10) above). These calculations were made with $k=h=0.05$;

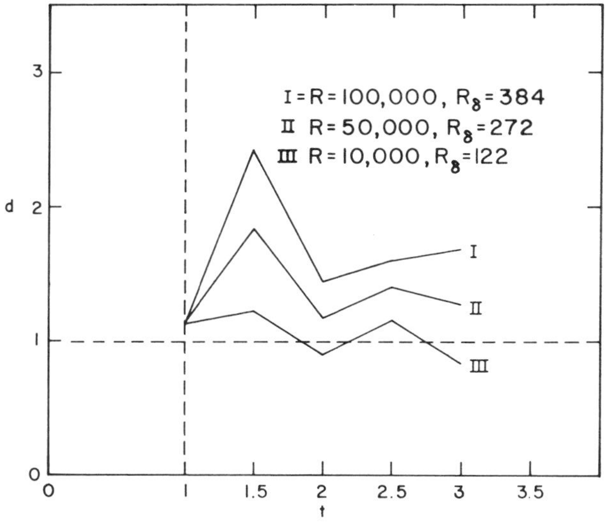

FIG. 2. *Growth of an unstable layer in two dimension.*

TABLE 1
Drag as a function of Reynolds number and time.

t	$R = 10000$ ($R\delta = 122$)	$R = 50000$ ($R\delta = 272$)	$R = 100000$ ($R\delta = 384$)
1	1.11	1.11	1.11
1.5	1.23	1.87	1.97
2	0.89	1.18	1.39
2.5	1.15	1.44	1.57
3	.77	1.25	1.65

the other parameters are as described in the preceding section: $\xi_{\max} = 0.6$, $X = 2$, $\sigma = h/\pi$. The point $X = \frac{1}{2}$ is in the middle of the region of interest. In our units, $\nu = R^{-1}$, and $R_\delta = \delta/\nu = 1.72\sqrt{R/2}$. From Table 1 and Fig. 2 one can see that $d(X)$ is growing for $R_\delta = 394$, $R = 10^5$; $d(X)$ is not growing for $R_\delta = 122$, $R = 10^4$, and $d(X)$ is initially excited but ultimately slowly decaying for $R_\delta = 272$, $R = 5 \times 10^4$. This last fact is debatable; the value $R_\delta = 272$ seems to be the approximate value of $R'_{\delta c}$. These results are reasonable in view of what is known from the theory and from experiments.

In Fig. 4 we exhibit the edge of the boundary layer as a function of x for $t = 3$, $R = 10^4$. The edge is defined as the smallest value of y for which $u = U_\infty$. The edge is not at infinity because we have finite number of vortex elements and thus the tail of the probability distribution of the locations of the elements is not accurately approximated. The layer is stable at this value of R, yet the edge is ragged and the layer appears to be "intermittent" (for a definition of intermittency see e.g. Cebeci and Smith [4]). The "intermittency" is due to the presence of discrete vortices; this connection will be

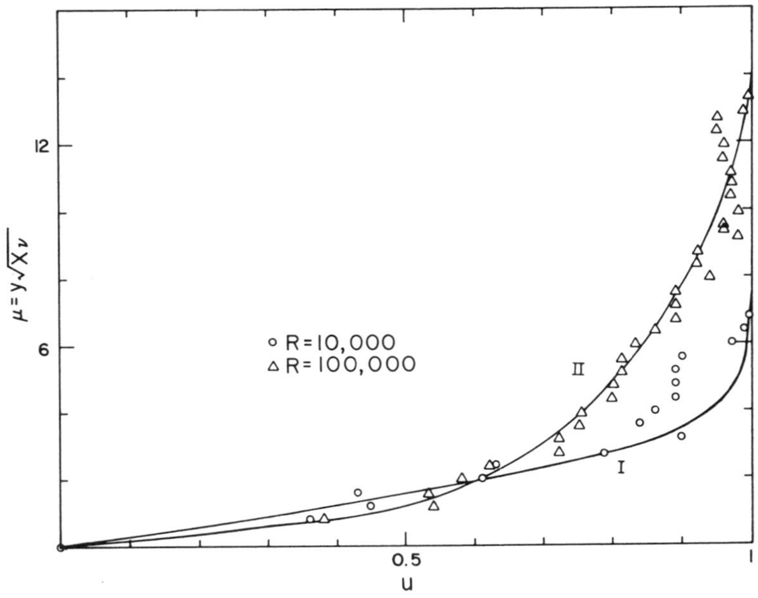

FIG. 3. *Velocity profiles in two dimensions.*

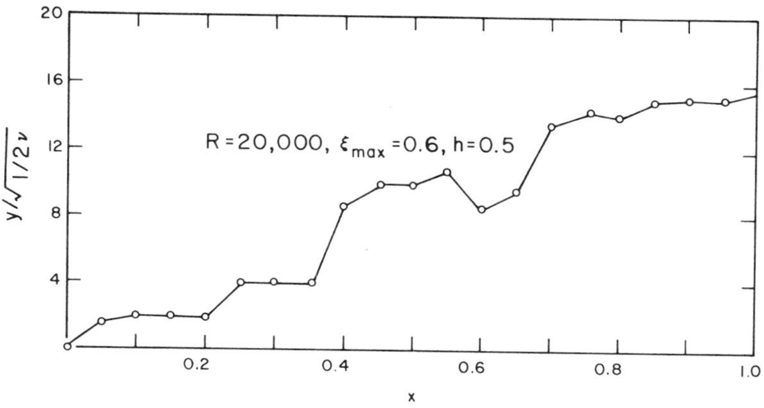

FIG. 4. *Boundary layer shape.*

exploited elsewhere for producing models of intermittency. It is obvious from Fig. 4 that the wave length of the growing modes cannot be determined directly from the instantaneous velocity distribution. However, it can be estimated indirectly. Consider the following question: how small must h be to allow us to distinguish between stable and unstable layers? Suppose that for $h > h_0$ this distinction can be made, but for $h \leqq h_0$ the layer appears to be stable even when it should not be. Then h_0 is an estimate of the wave length of the growing modes, since when $h \leqq h_0$ these modes are suppressed. In Table 2 we present the values of $d(X)$ at $X = \frac{1}{2}$ as a function of h for $R = 10^5$. We see that $10 < h_0 < 15$, in a reasonable if rough agreement with the Tollmien–Schlichting theory.

TABLE 2
Drag as a function of h, $R = 100000$, $R_\delta = 384$.

$h = 1/20$,	$k = 1/20$	$t = 1$	1.5	2	2.5	3
		$d = 1.11$	1.97	1.39	1.57	1.65
$h = 1/15$,	$k = 1/15$	$t = 1.27$	2	2.67	3.33	
		$d = 0.98$	1.48	1.66	1.70	
$h = 1/10$,	$k = 1/10$	$t = 1$	2	3		
		$d = 0.98$	1.10	1.08		

In Fig. 3 we display the velocity as a function of $\mu = y/\sqrt{\nu X}$ at $X = \frac{1}{2}$ for $R = 10^4$ and $R = 10^5$, averaged over 10 steps between $t = 2.5$ at $t = 3$. Curve I is the laminar steady Blasius profile, and curve II was drawn in what appears to the eye as a reasonable neighborhood of the points obtained at $R = 10^5$. The fluctuations are large (as one may well expect since $\xi_{max} = 0.6$), but the points at $R = 10^4$ are in a reasonable agreement with the Blasius curve; curve II (an unstable case) has a different shape. The gradients are first sharper, then smaller than in the stable case. This is consistent with experience in the unstable regime of thermal convection (see e.g. [6]). It is also consistent with data for a turbulent boundary layer in the following sense: The Tollmien–Schlichting waves are large scale structures in comparison with boundary layer thickness, while in the stable regime there are no organized structures. In the turbulent regime one can associate a velocity with an eddy size; the changes in the profile due to the transition from the stable to the unstable regime should be of the same nature as the changes in the velocity profile which occur when the eddy size increases. This is indeed the case (see Favre et al., [16]; their data are reproduced in Lighthill, [28]).

A typical run from $t=0$ to $t=3$ with the numerical parameters used here took about 10 minutes on the UC Berkeley CDC 6400 computer. At the end of the calculation, there were about 200 sheets and 300 blobs.

The physical problem in three space dimensions. We now consider the three dimensional version of the preceding problem.

Consider a semi-infinite flat plate placed on the half plane $z=0$, $x>0$. A fluid of density 1 occupies the half space $z>0$. At time $t<0$ the fluid is at rest, at $t=0$ the fluid is impulsively set into motion with velocity $U_\infty=1$. The Navier–Stokes equations in three dimensional space can be written in the form:

(11a) $$\partial_t \boldsymbol{\xi} + (\mathbf{u}\cdot\boldsymbol{\nabla})\boldsymbol{\xi} - (\boldsymbol{\xi}\cdot\boldsymbol{\nabla})\mathbf{u} = R^{-1}\Delta\boldsymbol{\xi},$$

(11b) $$\boldsymbol{\xi} = \operatorname{curl}\mathbf{u},$$

(11c) $$\operatorname{div}\mathbf{u} = 0.$$

$\mathbf{u}=(u, v, w)$ is the velocity vector, and $\mathbf{r}=(x, y, z)$ is the position vector. The boundary conditions are

(12a) $$\mathbf{u} = (U_\infty, 0, 0) \quad \text{for } z=\infty,\ t>0,$$

(12b) $$\mathbf{u} = 0 \quad \text{for } z=0,\ x>0,$$

(12c) $$\frac{\partial w}{\partial y} = 0 \quad \text{for } z=0,\ x<0.$$

Appropriate Prandtl equations can also be written. We shall need below only a simplified version of the equations, as well as the following fact about three dimensional boundary laminar layer approximations: The vertical component of the vorticity vanishes, i.e., for a solution of the Prandtl equations, $\boldsymbol{\xi}=(\xi_1, \xi_2, 0)$.

The Prandtl equations in three dimensions admit a two dimensional solution, the Blasius solution. That solution is unstable at high enough R. Squire's theorem (Lin, [29, p. 27]) states that the problem of instability to three dimensional infinitesimal perturbation is equivalent to a two dimensional problem at lower R.

Once the two dimensional perturbations begin to grow, several striking phenomena occur. In particular, before turbulence sets in, streamwise vortices (i.e. vortices whose axis is parallel to the mean flow) make their appearance. Intense secondary instabilities follow, and spots of intense motion emerge at random locations. An experimental investigation of boundary layer instability can be found in Klebanoff et al. [23]. Experimental investigations of turbulent boundary layers, in which phenomena resembling those which first arise immediately after the onset of instability persist and may be responsible for some of the observed features, are described e.g. in Favre et al. [16], Kline et al. [24], Willmarth [38]; theoretical aspects of several aspects of instability are found in Greenspan and Benney [17], Benney [3], Lighthill [28]. One of the major conclusions from the experimental data in Klebanoff et al., [23] is that the perturbed flow is periodic in the transverse direction (i.e., y direction). It is therefore natural to consider in three dimensions equations (11) with the added periodicity conditions

(13) $$\mathbf{u}(x, y+q, z) = \mathbf{u}(x, y, z), \qquad \boldsymbol{\xi}(x, y+q, z) = \boldsymbol{\xi}(x, y, z),$$

etc. Furthermore, from Klebanoff et al. (1962) we conclude that q is roughly equal to the streamwise wave length of the first unstable Tollmien–Schlichting waves; roughly, $q = 0.1$ in our units. We shall therefore be solving equations (11) with the boundary conditions (12) and (13), and $q = 0.1$.

The numerical methods in three dimensions. We consider first the three dimensional analogue of the blob method. The three dimensional problem is more difficult because the vorticity $\boldsymbol{\xi}$ is now a stretchable vector quantity which must satisfy div $\boldsymbol{\xi} = 0$.

In earlier three dimensional calculations (Leonard, [26], [27], Del Prete [13], Chorin, (unpublished)), the vorticity field was represented as a sum of vortex filaments. The difficulties with this approach are: (i) a huge amount of bookkeeping is required to keep track of the changing vortex configurations; (ii) there is no obvious way to generate the filaments at the boundary in a consistent manner. We bypass these difficulties by representing the vorticity as a sum of vortex segments (Fig. 5). Each vortex segment moves in the flow field induced by all the others. The condition div $\boldsymbol{\xi} = 0$ will be satisfied only approximately. The segments have no independent physical significance. The two dimensional blobs do not have one either; physical vortices or vortex tubes are expected to emerge from the superposition of the computational blobs or segments. A segment Λ is defined by seven quantities: The coordinates $\mathbf{r}^{(1)} = (x^{1)}, y^{(1)}, z^{(1)})$ of the center of its base, the coordinates $\mathbf{r}^{(2)} = (x^{(2)}, y^{(2)}, z^{(2)})$ of the center of its top, and its intensity κ. We shall write $\Lambda_i = (x_i^{(1)}, y_i^{(1)}, z_i^{(1)}, x_i^{(2)}, y_i^{(2)}, z_i^{(2)}, \kappa_i)$, $i = 1 \cdots, N$, N = number of segments. The base and the top are circles of radius σ, (the cut-off), which will be determined below.

Given a vorticity yield $\boldsymbol{\xi}(\mathbf{r})$, the velocity field in a fluid which fills out the whole space is given by the Biot–Savart formula (see e.g. [2]):

$$\mathbf{u}(\mathbf{r}) = -\frac{1}{4\pi}\int \frac{\mathbf{a}\times\boldsymbol{\xi}(\mathbf{r}')}{a^3}\,d\mathbf{r}' \tag{14}$$

$$\mathbf{a} = \mathbf{r} - \mathbf{r}', \qquad a = |\mathbf{a}|.$$

If the vorticity field is a sum of N closed vortex lines with the ith line having intensity κ_i, (14) becomes

$$\mathbf{u}(\mathbf{r}) = -\frac{1}{4\pi}\sum_{i\neq 1}^{N} \kappa_i \int_{i\text{th line}} \frac{\mathbf{a}\times\mathbf{s}}{a^3}\,ds. \tag{15}$$

$\mathbf{s} = \mathbf{s}(\mathbf{r}')$ is the unit tangent vector to the ith line at r', $ds = ds(r')$ is the arc length along the ith line, and as before $\mathbf{a} = \mathbf{r} - \mathbf{r}'$. We now seek an interaction law between vortex segments which will approximate the motion induced by (14) or (15).

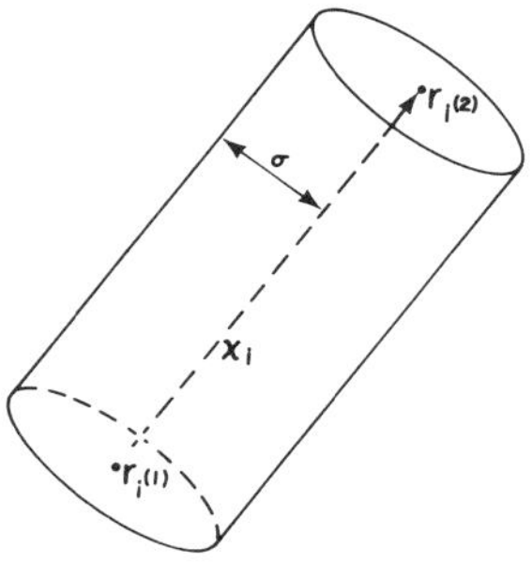

FIG. 5. *A vortex segment.*

Inside the segment the velocity field must be kept bounded, just as is the case in two dimension. Furthermore, the field must be modified inside the segments in such a way that the segments will be compatible with the boundary calculations (see below). The problem of the finding the correct formulation of the vortex method in three dimensions is difficult, (see e.g. Leonard [27]). The formulation offered here is plausible but not rigorously justified.

We require that the motion of a vortex ring or line made up of vortex segments should preserve the shape of the ring or line. This can be accomplished by ensuring that the configuration of the vectors **a** and of the velocity vectors which enter the formula for the motion of the tips of the segment is the appropriate translate of the corresponding configurations which determine the motion of the bases. Thus, let Λ_i, Λ_j be two vortex segments; define

$$\mathbf{r}_i^{(1)} = (x_i^{(1)}, y_i^{(1)}, z_i^{(1)}), \qquad \mathbf{r}_i^{(2)} = (x_i^{(2)}, y_i^{(2)}, z_i^{(2)}),$$

$$\mathbf{s}_j = \mathbf{r}_j^{(2)} - \mathbf{r}_j^{(1)},$$

$$\mathbf{a}_{ij}^{(1)} = \mathbf{r}_j^{(1)} - \mathbf{r}_i^{(1)},\ \mathbf{a}_{ij}^{(2)} = \mathbf{r}_j^{(2)} - \mathbf{r}_i^{(2)}, \quad \text{with } a_{ij}^{(1)} = |\mathbf{a}_{ij}^{(1)}|, \text{ etc.}$$

The velocity fields $\mathbf{G}_{ij}^{(1)}$, $\mathbf{G}_{ij}^{(2)}$ induced by Λ_j at $\mathbf{r}^{(1)}$ and $\mathbf{r}^{(2)}$ will be approximated by:

If $a_{ij}^{(1)} \geqq \sigma$ and $a_{ij}^{(2)} \geqq \sigma$:

(16a)
$$\mathbf{G}_{ij}^{(1)} = \frac{-\kappa_j}{4\pi} \frac{\mathbf{a}_{ij}^{(1)} \times \mathbf{s}_j}{(a_{ij}^{(1)})^3},$$

(16b)
$$\mathbf{G}_{ij}^{(2)} = \frac{-\kappa_j}{4\pi} \frac{\mathbf{a}_{ij}^{(2)} \times \mathbf{s}_j}{(a_{ij}^{(2)})^3}$$

If either $a_{ij}^{(1)} < \sigma$ or $a_{ij}^{(1)} < \sigma$:

(17a)
$$\mathbf{G}_{ij}^{(1)} = \frac{-\kappa_j}{4\pi} \frac{\mathbf{a}_{ij}^{(1)} \times \mathbf{s}_j}{\sigma^2 a_{ij}^{(1)}},$$

(17b)
$$\mathbf{G}_{ij}^{(2)} = \frac{-\kappa_j}{4\pi} \frac{\mathbf{a}_{ij}^{(2)} \times \mathbf{s}_j}{\sigma^2 a_{ij}^{(2)}}.$$

The equations of motion for each segment can now be obtained by summing the contributions of all the other segments and then adding to that sum the appropriate random component. This yields

(18a)
$$\mathbf{r}_i^{(1)n+1} = \mathbf{r}_i^{(1)n} + k \sum_{j \neq i} \mathbf{G}_{ij}^{(1)} + \boldsymbol{\eta},$$

(18b)
$$\mathbf{r}_i^{(2)n+1} = \mathbf{r}_i^{(2)n} + k \sum_{j \neq 1} \mathbf{G}_{ij}^{(2)} + \boldsymbol{\eta},$$

where $\mathbf{r}_i^{(1)n} \equiv \mathbf{r}_i^{(1)}(nk)$, etc., and $\boldsymbol{\eta}$ is a vector $\boldsymbol{\eta} = (\eta_1, \eta_2, \eta_3)$, with η_1, η_2, η_3, Gaussian random variable with means 0 and variances $2k/R$, independent of each other. $\boldsymbol{\eta}$ in (18a) is identical to $\boldsymbol{\eta}$ in (18b), since diffusion does not introduce rotation or stretching. The boundary condition at the wall can be satisfied as before by the introduction of appropriate image segments.

One can write the boundary layer equation in three dimensions and solve them by a method in which the computational elements are pieces of a vortex sheet (= "tiles") with sides h_1 in the x direction and h_2 in the y direction. Each tile carries a two dimensional vortex with components ξ_1, ξ_2. As observed earlier, $\xi_3 = 0$ in the boundary layer equations. However, we shall use the tiles only near the boundary, where vortex stretching is presumably negligible, or to create an initial Blasius profile, in which stretching is exactly zero. Therefore, the boundary layer equations we shall be solving reduce to

$$\partial_t \xi_1 + (\mathbf{u} \cdot \nabla)\xi_1 = \nu \frac{\partial^2 \xi_1}{\partial z^2}, \qquad \partial_t \xi_2 + (\mathbf{u} \cdot \nabla)\xi_2 = \nu \frac{\partial^2 z_2}{\partial z^2}$$

$$\xi_1 = \frac{\partial v}{\partial z}, \qquad \xi_2 = -\frac{\partial u}{\partial z}, \qquad \operatorname{div} \mathbf{u} = 0, \qquad \mathbf{u} = (u, v, w).$$

These equations can be solved by a straightforward extension of the sheet method described earlier. No vortex stretching will be taken into account, and we shall not take the trouble to write out the equations in full. The rejection and variance reduction techniques carry over from the two-dimensional case. Care is taken to ensure that $\sqrt{\xi_1^2 + \xi_2^2} \leqq \xi_{\max}$.

A tile created near the wall can become a segment if $t > T$ or if $z_i > l$. A segment which falls below l becomes a tile again. The transformation of tiles into segments (and vice versa) must obey the following conditions:

(i) A tile must become a segment parallel to the wall; i.e., if a tile $(x_i, y_i, z_i, \xi_{1i}, \xi_{2i})$ becomes a segment $(x_i^{(1)}, y_i^{(1)}, z_i^{(1)}, x_i^{(2)}, y_i^{(2)}, z_i^{(2)}, \kappa_i)$ we must have

(19a) $$z_i^{(2)} - z_i^{(2)} = 0.$$

(ii) A flow which is two dimensional when described by tiles must remain two dimensional when described by segments. The two dimensionality of a flow described by segments will be preserved only if the flow fields seen by the tips of the segments are translates of the flow fields seen by the bases, with a translation vector normal to the plane of the flow and pointing in the direction of a fixed normal $\mathbf{n}$ to that plane.

(iii) The stretching of the several segments represents the stretching of vorticity, which will be represented accurately only if the length of the segments is reasonably small. A reasonable normalization of that length in our problem is

(19b) $$y_i^{(2)} - y_i^{(1)} = h_2 \quad \text{when a segment is created.}$$

(iv) The circulation around a vortex line made up of tiles must equal the circulation around a vortex line made up of segments. If $y_i^{(2)} - y_i^{(1)}$ is normalized by (19b) this requirement leads to

(19c) $$\kappa_i = h_1 \sqrt{\xi_{1i}^2 + \xi_{2i}^2} \operatorname{sgn}(\boldsymbol{\xi}' \cdot \mathbf{n}),$$

where $\boldsymbol{\xi}' = (\xi_{1i}, \xi_{2i})$, $\mathbf{n}$ is the fixed normal to the plane of the flow and $\operatorname{sgn}(\alpha) = 1$ if $\alpha \geqq 0$, $\operatorname{sgn}(\alpha) = -1$ if $\alpha < 0$.

The remaining connecting formulas between segments and tiles are obviously

(19d) $$x_i^{(1)} = x_i,$$

(19e) $$y_i^{(1)} = y_i,$$

(19f) $$z_i^{(1)} = z_i.$$

Formulas (19) are of course invertible, and the computational elements can be treated as either tiles or segments, as the occasion warrants.

When two segments interact, their interaction is given by formulas (19); when a segment and a tile interact, they are both viewed as tiles.

Finally, the cut-off must be determined. We must require that if we consider on one hand the interaction of two infinite vortex lines parallel to the y axis represented by segments, and on the other hand the interaction of the same vortex lines represented by tiles, the former should approach the latter as the lines approach the wall. This requirement obviously reduces to the condition imposed on σ in two dimensions, and yields $\sigma = h_1/\pi$. This conclusion is of course legitimate only if most of the vorticity does indeed point in a direction parallel to the y axis.

Application of the numerical methods in three dimensions. In this section we discuss some of the features of the numerical method which are specific to the particular application at hand. Most of the numerical parameters are chosen just as they were chosen in the two-dimensional case; in particular, l and L. We picked $h_1 = k = \frac{1}{15}$, since the two-dimensional calculations showed that this was a minimal but adequate choice. We picked $h_2 = q/4$, after some experimentation showed that this value was sufficient to exhibit important effects.

The two major difficulties we encountered in three dimensions were: the large amount of computational labor, and the difficulty in imposing periodic boundary conditions on a grid-free method. The amount of labor is large not only because three-dimensional calculations are always more costly than two-dimensional calculations, but also (and especially) because the specific nature of the secondary instabilities which arise in three dimensions (see the next section) requires the creation of large amounts of vorticity at the walls. In consequence we used $\xi_{\max} = 1$. This value seems to yield results which are compatible with two-dimensional results obtained with smaller values of $\xi_{\max}$, but it is obviously so large that one may legitimately argue that what we have is a model rather than an approximation.

Periodic boundary conditions can be imposed on a vortex calculation, but the price in computing labor is high. There again we did the least we could reasonably do. For each vortex segment with base located at (x, y, z) (or its image created to satisfy the normal boundary condition, with a base at $(x, y, -z)$) we created two more segments, based at $(x, y \pm q, z)$, $q =$ the period and took their velocity fields into account when we moved the segment. Similarly, new tiles must be created outside the strip $0 \leqq y \leqq q$ with locations and strengths determined by periodicity. Some rather complex programming is needed to keep track of the several image systems as the tiles become segments and vice versa.

Finally, we note that if $\xi_1 = 0$ at $t = 0$, i.e., if there is no streamwise vorticity at all at $t = 0$, none will ever be created by our algorithm. Thus, if we are to observe the effects of streamwise vorticity, we must introduce some by artificial means. We proceeded as follows: At $t = 0$, for one time step, we changed the velocity at infinity. Instead of $\mathbf{u}(x, y, \infty) = (U_\infty, 0, 0)$ we set

$$\mathbf{u}(x, y, \infty) = \begin{cases} (U_\infty, A, 0) & \text{for } \frac{q}{4} < y < \frac{3q}{4}, \\ (U_\infty, 0, 0) & \text{elsewhere.} \end{cases}$$

We usually picked $A = 10^{-3}$ (note that $U_\infty = 1$). For $t > k$, we reverted to $\mathbf{u}(x, y, \infty) = (U_\infty, 0, 0)$ everywhere. The effect of this initial perturbation is to create a small streamwise vortex at the boundary, whose subsequent history is determined by diffusion, transport, and stretching.

Numerical results in three dimensions. Calculations done in three dimensions with $A = 0$ (i.e. with no perturbation which could trigger three dimensional effects) produce results similar to the results of two dimensional calculations. They afford a check on both, but are not worth discussing separately.

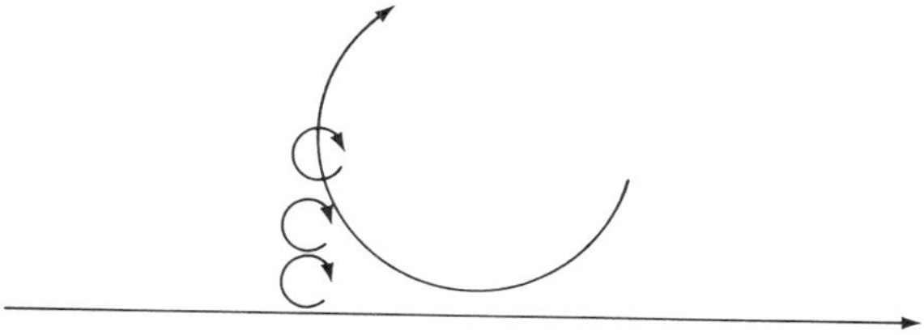

FIG. 6. *Amplification of streamwise rotation.*

Even a very small value of A (i.e. a very small three dimensional perturbation) has a substantial effect at all values of R we tried. The first phenomenon one observes when $A \neq 0$ is that the boundary layer becomes thicker than in the case $A = 0$. The mechanics of this effect are somewhat complex. A reasonable qualitative explanation is as follows: the rotation whose axis is parallel to the flow induces the creation of new streamwise vorticity at the wall. The new vorticity is then collected in streamwise strips in which the flow induced by the streamwise vortex leads away from the wall, while the regions where the induced flow points to the wall are depleted (see Fig. 6). The part of the boundary layer which thus expands can expand substantially, while the part which contracts cannot contract below zero. As a result the computed boundary layer thickness δ increases; δ at (X, Y) is defined by

$$\delta = \int_0^\infty (1 - u(x, y, \alpha)/U_\infty)\, d\alpha.$$

In Figs. 7 and 8 we plot the ratio δ/δ_b where δ = computed boundary layer thickness at $X = \frac{1}{2}$ averaged over a period in y, and δ_b = boundary layer thickness at $X = \frac{1}{2}$ computed from the steady Blasius solution. In Fig. 7, $R = 20000$. Note that at $t = 3$, R_δ computed with the steady δ is $R_\delta = 187$, and thus the layer should be steady. However, if R_δ is evaluated with the computed boundary layer thickness, R_δ at $X = \frac{1}{2}$ is approximately 300, and R_δ at $X = 1$ is approximately 440, well over the value at which the layer becomes unstable in the two-dimensional calculation. In Fig. 8, $R = 100000$,

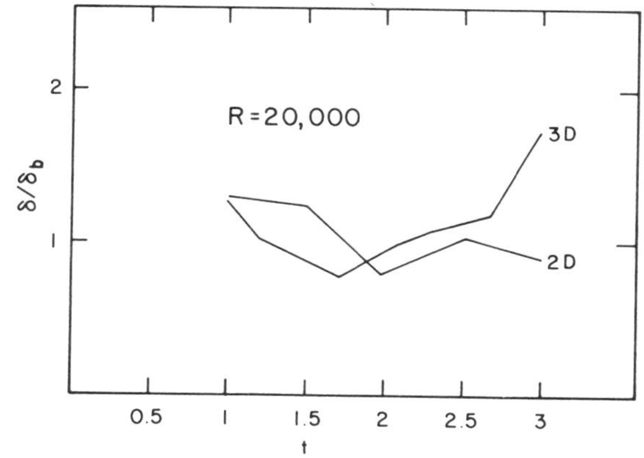

FIG. 7. *Growth of boundary layer thickness, $R = 20000$.*

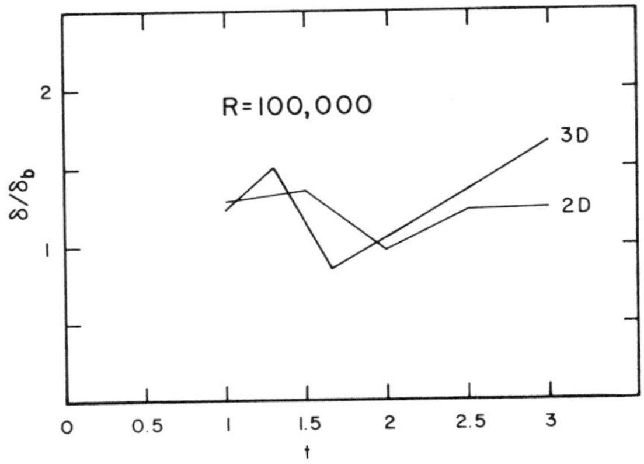

FIG. 8. *Growth of boundary layer thickness*, $R = 100000$.

and the same effect is reproduced. The computed values of the drag are not greatly affected by this thickening of the layer. (This is quite plausible, in view of the extra factor u in the integrand in the formula for the drag; the effect of this factor is to reduce the dependence of the drag on the velocity profile near the wall.)

When the layer becomes unstable to Tollmien–Schlichting waves, the streamwise vorticity begins to grow. The possible mechanisms for this growth are well known: The waves stretch lines; furthermore, they can create situations in which a horizontal streamwise line tilts away from the horizontal; its higher parts move faster than the lower parts, and stretching results. All segments are initially created with length h_2. If they stretch their length becomes $|\mathbf{r}_i^{(2)} - \mathbf{r}_i^{(1)}|$. The ratio $g = |\mathbf{r}_i^{(2)} - \mathbf{r}_i^{(1)}|/h_2$ is the stretching ratio. In Figs. 8 and 9 we plot $\bar{g}$, the average value of g, averaged over all segments. It is

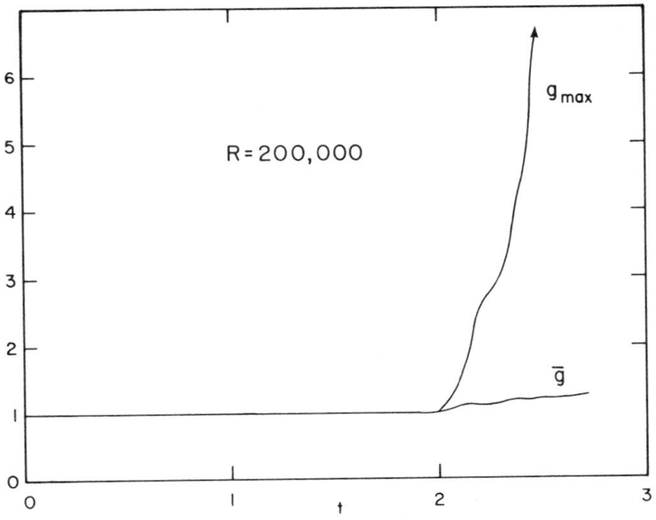

FIG. 9. *Amplification of boundary layer disturbances in three dimensions*, $R = 20000$.

seen to grow slowly with time. These figures are for $R = 20000$ and 100000. Note that at the time when $\bar{g}$ begins to grow with $R = 20000$ the layer had become thicker as a result of secondary motion and R_δ is larger than the critical value. In Fig. 9 we also plotted s, the total streamwise vorticity, and r, the ratio of newly created streamwise vorticity to newly created transverse vorticity. Roughly, r is an indication of the rate of growth of s. All these quantities are seen to grow slowly and steadily. The growth can be started earlier by increasing A. At value of R smaller than 10000, we never did succeed in inducing such growth within a time we could afford and without using very large values of A (i.e. A of order 1—not a plausible value for our problem).

The more interesting graph in Figs. 9 and 10 is the graph of the maximum value $g_{\max}$ of the stretching ratio. This value can become very large (~ 17), i.e., some vortices are stretched by a large amount. This suggests an extraordinary spottiness of the stretching process. This spottiness can be explained as follows: because our method is random, the local velocity profile can differ from point to point. At some points the local profile may be much more unstable than at others, and as a result secondary instabilities, whose growth rate is very large (Greenspan [17]) will occur at some points and not at others. One can also argue that as a result of the variation in local profiles, at some points the segments may depart from the horizontal more than at others, and therefore the stretching mechanism is more intense there. These two explanations may of course be identical. The "spots" make the major contribution to the growth of the mean quantities. Their presence indicates that the layer contains a mechanism for amplifying greatly small differences in local conditions. However, one should remember that our numerical layer is much noisier than a real layer is likely to be.

In Table 3 we display the values of the streamwise component of $\mathbf{u}$ at $x = ih_1$, $y = jh_2$, $z = 0$ and $R = 2 \times 10^5$, $t = 2.6$. The details of the fluctuations do not seem to have any particular physical significance. The values of R and t were picked somewhat arbitrarily; the table shows the spottiness of the field, and also shows that, as expected, the streamwise component of ξ increases as the layer thickens.

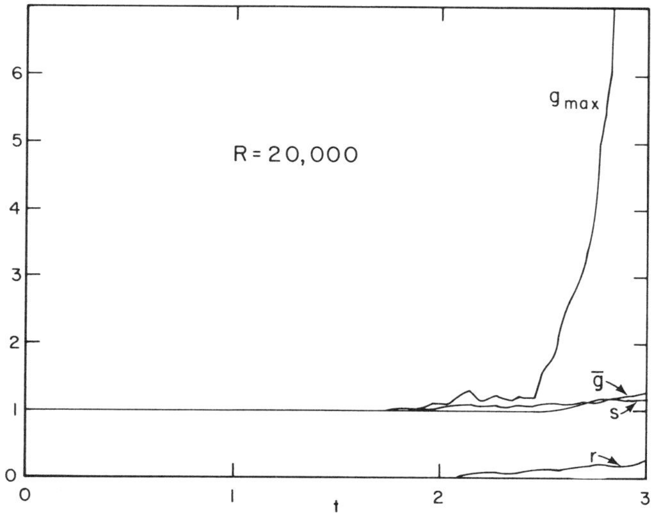

FIG. 10. *Amplification of boundary layer disturbances in three dimensions, $R = 100000$.*

When g increases, more and more segments and tiles have to be created; this is why we needed a large value of ξ_{max}. A further consequence is that computing for larger times than what we displayed is more expensive than we could afford. A typical run for $0 \leqq t \leqq 3$ took about one hour of CDC 6400 time at Berkeley.

TABLE 3
Streamwise vorticity at the boundary, $R = 2 \times 10^5$, $t = 2.6$

	$j=1$	2	3	4
$i=1$	.000	.000	.000	−.001
2	.000	.000	.000	.000
3	.000	.000	.000	.000
4	.003	.000	.000	.000
5	.028	.000	.001	.000
6	.000	.000	.000	.001
7	−.046	−.012	.001	.001
8	−.046	−.004	−.008	−.010
9	−.038	−.028	−.002	−.244
10	−.187	−.054	−.007	−.093
11	.091	−.065	.030	.076
12	−.062	−.310	.136	.205
13	2.260	.507	.480	1.070
14	−.077	.150	.826	−.621
15	−.552	.422	.001	−.273

Conclusions. Our vortex methods, including the new three dimensional version and the new vorticity creation procedure, seem to be able to reproduce important features of boundary layer behavior in two and three dimensions and at Reynolds numbers where instability is expected. The three-dimensional calculation does exhibit a growth of streamwise vorticity as well as spottiness; however, it was not performed for times long enough for anything resembling fully developed turbulence to be present. Unlike other methods, our methods are not limited at high R by the difficulty in distinguishing real from numerical diffusion; they are however limited, like other methods, by the fact that effects not resolved cannot be seen; i.e., if there are not enough computational elements to represent a phenomenon, that phenomenon will not be observed. Since fully turbulent flow is very complicated, our methods do not remove the need for careful modeling in some practical applications.

REFERENCES

[1] W. ASHURST, *Numerical simulation of turbulent mixing layer dynamics*, SANDIA (Livermore) Report (1977).
[2] G. K. BATCHELOR, *An Introduction to Fluid Mechanics*, Cambridge University Press, London, 1967.
[3] D. J. BENNEY, *A nonlinear theory for oscillations in a parallel flow*, J. Fluid Mech., 10 (1960), p. 209.
[4] T. CEBECI AND A. M. O. SMITH, *Analysis of Turbulent Boundary Layers*, Academic Press, New York, 1974.
[5] A. CHEER, *Program BOUNDL*, LBL-6443 Suppl. Report, Lawrence Berkeley Lab. (1978).
[6] A. J. CHORIN, *A numerical method for solving incompressible flow problems*, J. Comput. Physics, (1967), p. 12.
[7] ———, *Numerical study of slightly viscous flow*, J. Fluid Mech., 57 (1973), p. 785.
[8] ———, *Gaussian fields and random flow*, Ibid., 63 (1974), p. 21.
[9] ———, *Lectures on Turbulence Theory*, Publish/Perish, Boston (1976).
[10] ———, *Vortex sheet approximation of boundary layers*, J. Comput. Physics, 27 (1978), 428.

[11] A. J. CHORIN AND J. E. MARSDEN, *A Mathematical Introduction to Fluid Mechanics*, Springer Verlag, New York, 1979.
[12] A. J. CHORIN, T. J. R. HUGHES, M. F. MCCRACKEN AND J. E. MARSDEN, *Product formulas and numerical algorithms*, Comm. Pure Appl. Math., 31 (1978), p. 205.
[13] V. DEL PRETE, *Numerical simulation of vortex breakdown*, LBL Math. & Comp. Report 1978.
[14] W. ECKHAUS, *Studies in Nonlinear Stability Theory*, Springer Verlag, New York, 1965.
[15] H. FASEL, *Investigation of the stability of boundary layers in a finite difference model of the Navier–Stokes equations*, J. Fluid Mech., 78 (1970), p. 355.
[16] A. FAVRE, J. GARIGLIO AND J. DUMAS, Phys. Fluids, 12, Suppl. (1967), p. 138.
[17] H. P. GREENSPAN AND D. J. BENNEY, *On shear layer instability, breakdown and transition*, J. Fluid Mech., 15 (1963), p. 133.
[18] O. HALD, *The convergence of vortex methods II*, SIAM J. Numer. Anal., 16 (1979), p. 726.
[19] O. HALD AND V. DEL PRETE, *The convergence of vortex methods*, Math. Comp., 32 (1978), p. 791.
[20] J. M. HAMMERSLEY AND D. C. HANDSCOMB, *Monte Carlo Methods*, Methuen, 1964.
[21] R. JORDINSON, *The flat boundary layer, Part I. Numerical integration of the Orr–Sommerfeld equation*, J. Fluid Mech., 43 (1970), p. 801.
[22] H. T. KIM, S. J. KLINE AND W. C. REYNOLDS, *The production of turbulence near a smooth wall in a turbulent boundary layer*, Ibid., 50 (1971), p. 133.
[23] P. S. KLEBANOFF, K. D. TIDSTROM AND L. D. SARGENT, *The three dimensional nature of boundary layer instability*, Ibid., 12 (1962), p. 1.
[24] S. J. KLINE, W. C. REYNOLDS, F. A. SCHRAUB AND P. W. RUNDSTADLER, *The structure of turbulent boundary layers*, Ibid., 30 (1967), p. 741.
[25] J. LAMPERTI, *Probability Theory*, Benjamin, New York, 1966.
[26] A. LEONARD, *Numerical simulation of interacting, three dimensional vortex filaments*, Proc. 4th Int. Conf. Num. Methods Fluid Dynamics, Springer Verlag, New York, 1975.
[27] ———, *Simulation of unsteady three dimensional separated flows with interacting vortex filaments*, Proc. 5th Int. Conf. Num. Mech. Fluid Dynamics, Springer Verlag, New York, 1977.
[28] M. J. LIGHTHILL, *Turbulence*, Osborne Reynolds and Engineering Science Today, Manchester Univ. Press, 1976.
[29] C. C. LIN, *The Theory of Hydrodynamic Stability*, Cambridge University Press, London, 1966.
[30] M. F. MCCRACKEN AND C. S. PESKIN, *The vortex method applied to blood flow through heart valves*, Proc. 6th Int. Conf. Num. Methods Fluid Dynamics, Springer Verlag, New York, 1978.
[31] D. MEKSYN AND J. T. STUART, *Nonlinear instability*, Proc. Roy. Soc. London, A, 208 (1951), p. 517.
[32] L. ONSAGER, *Statistical hydrodynamics*, Nuovo Cimento, 6, Suppl. (1949), p. 229.
[33] E. RESHOTKO, *Boundary layer stability and transition*, Ann. Rev. Fluid Mech., 8 (1976), p. 311.
[34] H. L. ROGLER AND E. RESHOTKO, *Disturbances in a boundary layer introduced by a low intensity array of vortices*, SIAM J. Appl. Math., 28 (1975), p. 431.
[35] H. SCHLICHTING, *Boundary Layer Theory*, McGraw-Hill, New York, 1960.
[36] A. I. SHESTAKOV, *A hybrid vortex—ADI solution for flows of low viscosity*, J. Comput. Phys., 31 (1979), p. 313.
[37] A. A. TOWNSEND, *Boundary Layer Research*, Freiburg Symposium, H. Gortler, Ed., Springer Verlag, 1958.
[38] W. W. WILLMARTH, *Structure of turbulence in boundary layers*, Advances in Appl. Mech., 15 (1976), p. 159.

NUMERICAL METHODS FOR USE IN COMBUSTION MODELING

Alexandre Joel Chorin*
Department of Mathematics
University of California
Berkeley, California

In recent years a number of methods have been developed for use in combustion modeling; some of these methods rely in an essential way on a random walk or a sampling procedure. The purpose of this talk is to explain in an elementary manner how these methods work and why they are useful.

Combustion problems have the following characteristics, each one of which presents a major challenge to computational modeling: they are time dependent phenomena occuring in a three dimensional space; the number of equations to be solved is often substantial; there are several distinct length and time scales to be resolved; the flow in which the combustion occurs is usually turbulent, and on the scales important to combustion, very intermittent. The solution is very sensitive to a numerically induced diffusion or conduction.

Stochastic methods provide a particularly promising way for overcoming these difficulties. They are typically insensitive to stiffness; they provide a way of controlling numerical diffusion (and thus of ensuring that the effects of viscosity or conduction are represented accurately), and they concentrate computing effort where it would do the most good. We shall explain the methods by means of simple examples.

A Reaction-Diffusion Equation

Consider the equation

$$v_t = \kappa v_{xx} + f(v) \tag{1}$$

where κ is a small parameter, and $f(v)$ is a function of v such that: $f(v) = 0$ for $0 \leq v \leq \varepsilon$; $f(v) > 0$ for $\varepsilon < v < 1$, $f(1) = 0$, $f'(1) < 0$. Equation (1) is a model for many equations which occur in combustion theory. Consider in particular the initial data

* Partially supported by the Engineering, Mathematical, and Geosciences Division of the U.S. Department of Energy under contract W-7405-ENG-48, and by the Office of Naval Research under contract N00014-76-0-0316.

Reprinted with permission from *Computing Methods in Applied Sciences and Engineering*, No. 4, 1980.

$$v(x,0) = \begin{cases} 1 & x < 0, \\ 0 & x > 0. \end{cases} \tag{2}$$

The condition $f(v) = 0$ for $0 \le v \le \varepsilon$ (ε is an "ignition temperature") makes the problem well conditioned and allows a simple discussion (see e.g. [4], [10]).

Equation (1) has a solution which tends to a traveling wave solution. Both the thickness and the velocity of the wave are $O(\sqrt{\kappa})$. Suppose one were to compute v by a difference method with a fixed grid size. Clearly, the amount of labor required to reach a fixed time T with a given accuracy would be $O(\kappa^{-\frac{1}{2}})$. What is needed is a procedure for refining the mesh in the region where v has large gradients, without disturbing the rest of the calculation (Equation (1) may be describing the flow in a small area only) and without incurring an instability. We wish to show that a random procedure will do the job perfectly, in the sense that the amount of labor for given accuracy and given integration time will be exactly independent of κ. The algorithm will be explicit and unconditionally stable.

Approximate $v(x,t)$ by a piecewise constant function of x, with jumps Δv_i, $i = 1, \ldots, N$ at the points x_i, $i = 1, \ldots, N$. Assume $|\Delta v_i| < \Delta_{max}$ for all i, Δ_{max} a small quantity. Assume for simplicity that $v(x = +\infty, t) = 0$. Let the time be descretized, $t = nk$, n = integer, $n \ge 0$, k = time step. Then

$$v(x,nk) = \sum_{x_i > x} \Delta v_i \tag{3}$$

It is not required that the x_i be distinct. Thus, if $\Delta_{max} = 1/N$, the initial data (2) can be represented by

$$\Delta v_1 = \ldots = \Delta v_N = 1/N,$$

$$x_1 = \ldots = x_N = 0.$$

To progress from $t = nk$ to $t = (n+1)k$, let each of the x_i perform a gaussian jump,

$$x_i^{n+1} = x_i^n + \eta ,$$

where η is drawn from a gaussian distribution with mean zero and variance $2\kappa k$. The expected value of (3) is then a solution of the heat equation $v_t = \kappa v_{xx}$. Note that what is diffused by a random walk is not v itself, but its x derivative. This is the key to success.

We have not yet taken $f(v)$ into account. Consider a region S_i

in which v is a constant, $S_i = \{x | x_i \le x \le x_{i+1}\}$. $v = v_i$. Solve the ordinary differential equation

$$\frac{dw}{dt} = f(w), \quad w(0) = v_i.$$

Let $\delta_i = w(k) - v_i$ be the change in v due to f in S_i. Then v_i should become $v_i + \delta_i$ in S_i, and thus Δv_{i+1} becomes $\Delta v_{i+1} + \delta_i$ and v_i becomes $\Delta v_i - \delta_i$. Do this for all S_i. In after this step one of the $|\Delta v_i|$ exceeds Δ_{max}, break it into smaller pieces which do satisfy $|\Delta v_i| \le \Delta_{max}$. If a Δv_i becomes very small, delete it (some care is required to ensure that $v(x = -\infty, t) = 1$ with the initial data (2)). This last step ensures that no work is performed where the gradients of v are small. The overall algorithm is easy to program and use, and with appropriate variance reduction it is also very accurate. Furthermore, as $\kappa \to 0$, the solution scales exactly with $\sqrt{\kappa}$ and the amount of labor remains constant. For details see [4].

The Boundary Layer Equations

The boundary layer equations for an incompressible fluid are conceptually similar to a reaction diffusion equation. They describe the interaction between the creation of vorticity at a wall, its diffusion and its transport. The creation process is more interesting than in a reaction-diffusion equation.

Consider a flow over a heat plate in two dimensions. The Prandtl equations are

$$\partial_t \xi + (\underline{u} \cdot \underline{\nabla})\xi = \nu \frac{\partial^2 \xi}{\partial y^2}, \tag{4a}$$

$$\partial_x u + \partial_y v = 0, \tag{4b}$$

$$\frac{\partial u}{\partial y} = -\xi, \tag{4c}$$

where $\underline{u} = (u,v)$ is the velocity, ξ is the vorticity, x is a coordinate parallel to the wall, y is a coordinate normal to the wall, and ν is a (small) viscosity. The boundary conditions include: $u(x,y = +\infty, t) = U$, $\underline{u}(x,y = 0,t) = 0$. We shall solve the equations by diffusing ξ, which by (4c) is a velocity gradient.

Consider a collection of N segments S_i of vortex sheets. Each S_i, $i = 1, \ldots, N$, is a segment of a straight line, parallel to the wall, of length h, with center $\underline{x}_i = (x_i, y_i)$, and such that if $u+$ is u just above S_i and $u-$ is u just below

S_i, $u+ - u- = \xi_i$. Consider the motion of the S_i as equations (4) would impose it. Equations (4b) and (4c) yield

$$u(x,y) = U - \int_y \xi(x,z)\,dz, \tag{5a}$$

$$v(x,y) = -\partial_x \int_0^y u(x,z)\,dz\ , \tag{5b}$$

i.e., given $\xi(x,y)$, u and v can be determined. Equation (5a) states that the fluid element at (x,y) is slowed down from U by all the sheets above that element. Equation (5b) states that whatever fluid cannot leave a rectangular region through its sides must leave upwards. These observations lead quickly to rules which determine $\underline{u}_i = \underline{u}(x_i)$ given the array of $S_j, j = 1, \ldots, N$, and one can write approximately

$$u_i^{n+1} = u_i^n + ku_i \tag{6b}$$

$$y_i^{n+1} = y_i^n + kv_i \tag{6b}$$

where $t = nk$, n = integer, k = time step, as before. Again, the diffusive term in (4a) can be taken into account by writing, instead of (6b),

$$y_i^{n+1} = y_i^n + kv_i + \eta, \tag{7}$$

where η is a gaussian random variable with mean 0 and variance $2k\nu$. (5a) and (5b) clearly lead to rules which satisfy the condition $u = U$ at $y = +\infty$ at $v = 0$ at $y = 0$.

Suppose the boundary condition $u = 0$ fails to be satisfied at a point $(x,0)$ on the wall. Imagine the flow is continued from $y > 0$ to $y < 0$ by the rule $\underline{u}(x,-y) = -\underline{u}(x,y)$. As one crosses the wall downwards, both u and y change signs, and therefore

$$\xi(x,y) = -\frac{\partial u}{\partial y}(x,y) = -\frac{\partial u}{\partial y}(x,-y) = \xi(x,-y)\ .$$

Now let the flow described by equation (4a) take place. The diffusion term will guarantee that the boundary condition $u = 0$ is satisfied.

It u is not zero at the wall, the antisymmetry $u(x,-y) = -u(x,y)$ will introduce at the wall a line of discontinuity. This line of discontinuity can be broken into vortex sheets of the type described above, which then take part in the subsequent motion. This is the process of vorticity generation, which mimics the physical process of vorticity generation. For details, implementation, and variance reduction, see e.g. [3], [5]. The algorithm is unconditionally stable, and contains an automatic scaling which scales the thickness of the layer with $\sqrt{\nu}$.

Vortex Method for the Navier-Stokes Equations

The step from the Prandtl equations to the Navier-Stokes equation is short. In the plane, the latter equations can be written as

$$\partial_t \xi + (\underline{u}\cdot\underline{\nabla})\xi = \frac{1}{R}\Delta\xi, \tag{8a}$$

$$u = -\partial_x\psi, \quad v = \partial_y\psi, \tag{8b,c}$$

$$\Delta\psi = -\xi, \quad \underline{u} = (u,v) \ , \tag{8d}$$

where u, v, ξ have the same meaning as in (4). ψ is the stream function and R is the Reynolds number (which we assume is large). The main difference between these quations and the Prandtl equations lies in the fact that the relationship between $\underline{u}$ and ξ is more complex, and also in the fact that diffusion takes place in both spatial direction. Equation (8a) states, just as does equation (4a), that in the absence of diffusion vorticity is merely transported by the velocity field which itself induces.

What should be done is plain: consider ξ to be a sum of small elements. Move each according to the velocity field induced by all the others. Add a small random component to represent diffusion. Create vorticity at walls. For details, see [1], [5], [11].

Two words of caution are needed: at the wall, $\underline{u}(x,y) = -\underline{u}(x,-y)$ does not lead to $\xi(x,y) = \xi(x,-y)$ for the Navier-Stokes equations. The reason is that here,

$$\xi = -\frac{\partial u}{\partial y} + \frac{\partial v}{\partial x}$$

and though u, v and y change signs as one crosses the wall, x does not (we assume as before that the wall coincides with $y = 0$). This leads to some minor tehcnical difficulties. The easiest thing to do is use sheets near walls.

Finally, note that the cost of counting all vortex interactions can be substantial. One may be tempted to impose a grid on the domain of interaction, extrapolate the vorticity to the grid, compute the velocity from (8b,c,d) on the grid by some fast method, and interpolate to the vorticity elements (this is the "vortex in cell" or "vortex cloud" method). The idea is attractive but dangerous. Interpolation introduces a numerical viscosity, which, if one is careless, may destroy the purpose of the method, which is to represent the diffusive length scales correctly. In many problems, of course, the diffusion does not matter as long as it is small (e.g. in a periodic domain or in the absence of boundaries) but then one may just as well compute with more standard methods.

Glimm's Method for Reacting Gas Flow

We now turn to a somewhat different order of ideas, which goes back to Glimm's construction of solutions of hyperbolic problems [8].

Consider the simplicity the single equation

$$v_t = f_x, \quad f = f(v) \tag{9}$$

A Riemann problem for this equation is the initial value problems with data

$$v(x,0) = \begin{cases} v_R & x > 0\ , \\ v_L & x < 0\ . \end{cases}$$

v_R, v_L constants. Let the solution of this problem be $w(v_R, v_L, x,t)$. Consider now the initial value problem for (9) with data $v(x,0) = g(x)$. Let $x = ih$, i = integer, h = spatial increment (we are now for the first time introducing a spatial grid). Let $t = nk$. Given v_i^n, v_{i+1}^n, we set

$$v_{i+1/2}^{n+1/2} = w(v_{i+1}^n, v_i^n, \theta h, k/2),$$

where θ is a variable whose values are equidistributed on $[-\frac{1}{2}, \frac{1}{2}]$. This formula defines an algorithm for solving (9) on a staggered grid. The algorithm proceeds by constructing exact solutions of the Riemann problems, and then sampling them. In the simple case $f(v) = v$, the equation to be solved becomes $v_t = v_x$; the solution is $v(x,t) = g(x+t)$. The Glimm scheme reduces to

$$v_{i+1/2}^{n+1/2} = \begin{cases} v_{i+1}^n & \text{if} \quad \theta h \geq -k/2 \\ v_i^n & \text{if} \quad \theta h < -k/2. \end{cases}$$

A brief recursion argument shows that

$$v_i^n = g(x+\phi,\ t),$$

where ϕ is a variable whose mean is zero and whose variance tends to zero as $h \to 0$, $k/h \leq 1$. ϕ is independent of x. The Glimm scheme for this simple equation preserves the shape of the solution exactly, at the cost of a small random displacement.

We adopted the Glimm scheme in a combustion model because it had a small intrinsic diffusion. (In two dimensional problems it turns out that the diffusion is small but not zero). This is particularly significant in the following circumstances: consider a flow in which thin combustion layers are imbedded. On the natural outer scale one may often view the diffusion and conduction coefficients as being zero and the reaction rates as infinite. The equations of gas dynamics acquire then many of the properties of non convex hyperbolic systems (see e.g. [2], [5]). A simple non convex hyperbolic equation is an equation (9) in which f'' is allowed to change signs. In such a problem, the wave pattern which emerges depends on the global shape of the function f, and diffusion in a rather substantial amount is needed to spread the variation of the solution sufficiently for a difference scheme to discern how it should behave (see [9]). The Glimm scheme overcomes this problem entirely by building the right wave patterns into the Riemann problems, without a need for extra diffusion. For examples, see [6], [7], [11].

Conclusions

The real problems of combustion theory are of course much more complex than the simple examples sketched above, and the actual programs in use are hybrids of substantial complexity. All the methods described above have in common the fact that they rely heavily on known properties of the specific equations to be solved. In very complicated problems, this may well be the right path to choose.

BIBLIOGRAPHY

[1] A.J. Chorin, J. Fluid Mech., 57 785 (1973).

[2] A.J. Chorin, J. Comp. Phys., 25, 257 (1977).

[3] A.J. Chorin, J. Comp. Phys., 27 428 (1978).

[4] A.J. Chorin, Stochastic Solution of Reaction/Diffusion Equations, Berkeley (1979).

[5] A.J. Chorin and J.E. Marsden, A Mathematical Introduction to Fluid Mechanics, Springer (1979)

[6] P. Colella, Ph.D. Thesis, Dept. of Math, Berkeley (1979).

[7] P. Concus and W. Proskurowski, J. Comp. Phys., 30, 153.(1979).

[8] J. Glimm, Comm. Pure Appl. Math, 18, 697 (1965).

[9] A. Harten, J. Hyman and P.D. Lax, Comm. Pure Appl. Math, 29, 297 (1976).

[10] S.S. Lin, Ph.D. Thesis, Dept. of Math, Berkeley (1975).

[11] A. Majda, A Qualitative Model for Dynamic Combustion, Berkeley, Dept. of Math (1979).

[12] A. Sod, Numerical Methods in Fluid Mechanics, Academic Press (1980).

NUMERICAL MODELLING OF TURBULENT FLOW IN A COMBUSTION TUNNEL

By A. F. GHONIEM, A. J. CHORIN and A. K. OPPENHEIM
Lawrence Berkeley Laboratory, University of California, Berkeley, California 94720, *U.S.A.*

(*Communicated by H. Jones, F.R.S. – Received* 25 *February* 1981)

[Plate 1]

CONTENTS

A numerical technique is presented for the analysis of turbulent flow associated with combustion. The technique uses Chorin's random vortex method (r.v.m.), an algorithm capable of tracing the action of elementary turbulent eddies and their cumulative effects without imposing any restriction upon their motion. In the past, the r.v.m. has been used with success to treat non-reacting turbulent flows, revealing in particular the mechanics of large-scale flow patterns, the so-called coherent structures. Introduced here is a flame propagation algorithm, also developed by Chorin, in conjunction with volume sources modelling the mechanical effects of the exothermic process of combustion. As an illustration of its use, the technique is applied to flow in a combustion tunnel where the flame is stabilized by a back-facing step. Solutions for both non-reacting and reacting flow fields are obtained. Although these solutions are restricted by a set of far-reaching idealizations, they nonetheless mimic quite satisfactorily the essential features of turbulent combustion in a lean propane–air mixture that were observed in the laboratory by means of high speed schlieren photography.

Introduction

Numerical analysis of turbulent flow has traditionally been based on some form of finite-difference treatment of appropriately averaged Navier–Stokes equations, supplemented by an adequate set of relations to correlate the turbulent flow parameters: the closure model.

20-2

Reprinted with permission from *Philosophical Transactions of the Royal Society of London, Series A*, Vol. 304, 1982.

For this purpose it is customary to apply first the Reynolds splitting principle to all the dependent variables. Each term in the governing equations is then appropriately averaged. This may involve either time, or ensemble, or Favre-mass averaging, depending on whether one is seeking a steady-state, or time-dependent, or compressible-flow solution. Owing to the nonlinearity of these equations, double correlations of the fluctuating components arise, while the averaging process involves essentially an integration, as a consequence of which a loss of information is incurred. The usual way to remedy this is to introduce a system of relations between the correlations and some mean flow parameters: the closure relations. To obtain a numerical solution, a finite-difference technique is then applied, yielding the description of the flow field in terms of discrete values of its parameters on the nodes of an Eulerian mesh.

The averaging-closure-differencing method (a.c.d.m.) described above has been used in many variations, producing satisfactory results in good agreement with experimental data for a wide assortment of turbulent flow problems. Work in this field has been reviewed recently by Mellor (1979), McDonald (1979), and, with particular reference to modern methods based on the use of the probability density function (p.d.f.), Williams & Libby (1980). These reviews have demonstrated the value of a.c.d.m. as a powerful analytic tool for the study of turbulent combustion.

The a.c.d.m., however, is handicapped by several drawbacks. The following are particularly relevant to the problem at hand.

(i) The averaging process deprives the equations of essential information about the mechanism of turbulence; this necessitates introducing turbulence models on heuristic grounds rather than obtaining information about them from the solution.

(ii) The turbulence model required for the closure relations has to be postulated and the value of its parameters has to be adjusted to match experimental data.

(iii) The finite-difference technique introduces numerical diffusion, which tends to smooth out local perturbations, an effect that is especially harmful at high Reynolds numbers, where regions of substantial shear arise in the flow field; the effect of the a.c.d.m. in this respect is to curtail the Reynolds number, causing the misrepresentation of some of the most essential features of the flow field.

(iv) The effects of exothermic processes of combustion on the flow field are particularly difficult to handle by the a.c.d.m.; as pointed out over five years ago by Williams (1974), these processes cause many fold increases in specific volume and occur at rates that are so high that taking them properly into account in a finite-difference scheme is associated with practically insurmountable difficulties.

All these drawbacks are addressed by the random vortex method (r.v.m.) developed by Chorin (1973). This method was designed to develop a satisfactory approximation to the solution without finite-difference treatment of the equations. Essential features of the flow field governed by the Navier–Stokes equations are mimicked by the action of vortex elements that model the essential ingredients of turbulence, the elementary eddies. Their random walks express the effects of diffusion, while compliance with the tangential boundary condition at the walls is assured by creation of the proper amount of vorticity. A potential flow solution is used at the same time in accordance with the principle of fractional steps to guarantee that the normal boundary condition is satisfied.

The r.v.m. keeps track of the position and strength of all the vortex elements constituting the flow field and is thus essentially grid-less. It is therefore devoid of the smoothing, intrinsic to the

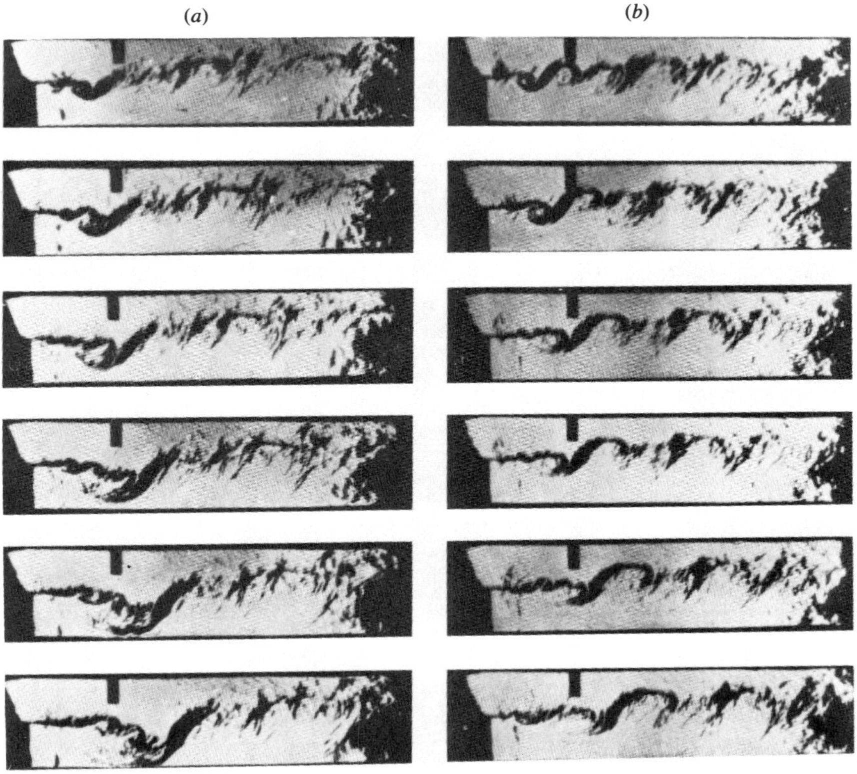

FIGURE 1. Photographic schlieren records of turbulent combustion stabilized behind a step, in a propane-air mixture at an equivalence ratio of 0.57, entering the channel at $u'_{\infty} = 13.6$ m s^{-1} ($R = 22 \times 10^4$) while $T_{\infty} = 295$ K: (*a*) growth of a large eddy under the influence of recirculation (time interval between frames: 1.22 ms); (*b*) 'steady-state' propagation of a large-scale ('coherent') structure (time interval between frames: 1.16 ms).

finite-difference technique, and unaffected by the numerical diffusion it introduces. Above all, the r.v.m. does not involve any averaging whatsoever. On the contrary, instead of damping the disturbances, it actually introduces some randomness, or numerical noise, simulating the mechanism of local perturbations in a real flow.

Partial convergence proofs for the r.v.m. have been provided by Chorin *et al.* (1978) and Hald (1979). In particular, the error in the solution was shown to be proportional to the inverse of the square root of the Reynolds number, furnishing further evidence of the eminent suitability of the r.v.m. to the analysis of turbulent flows. Its success in this respect has been amply demonstrated by solutions obtained for flows around solid bodies (Chorin 1973; Cheer 1979), shear layer effects (Ashurst 1979, 1981) and internal flows (McCracken & Peskin 1980). As pointed out by Roshko (1976), it was indeed instrumental in revealing the mechanics of large-scale turbulence patterns, the so-called 'coherent structure', by elucidating such features as the shear layer mechanism and the processes of eddy shedding, growth, intertwining, and pairing.

The most prominent aspects of the r.v.m. are presented here from an entirely pragmatic point of view. The algorithm, augmented to accommodate the effects of flames, is then applied to the analysis of turbulent flow with combustion stabilized in the recirculation zone behind a step.

Salient features of such a flow field are displayed in figure 1 (plate 1), a selection of photographic schlieren records presented by Ganji & Sawyer (1980). The large-scale vortex pattern characteristic of the 'coherent structure' is clearly discernible, while the flame front is recorded by dark streaks, the loci of maximum gradient in the refractive index reflecting the rapid change in density and temperature due to combustion. The records were obtained for a propane–air mixture at an equivalence ratio of 0.57, initial temperature of 295 K flowing at a velocity of 13.6 m s^{-1} (Reynolds number 2.2×10^4) in the inlet channel, 2.54 cm wide, into a test section 5.08 cm wide and 17.3 cm deep. There are two sequential series made out of extracts from the same high speed film. Series (*a*) shows the process of the coalescence of eddies and their intrusion into the recirculation zone, the time interval between frames being 1.22 ms. Series (*b*) shows the normal formation and development of eddies in the mixing zone, the time interval between frames being 1.16 ms.

The analysis is restricted by a formidable array of simplifying idealizations. The most important among them is the restriction to two-dimensional flow fields, as a consequence of which many interesting features of turbulence cannot be treated. It should be stressed, however, that this is not tantamount to limitations of the r.v.m. itself. In fact it has been recently extended and applied to the study of boundary layer transition in a three-dimensional flow field (Chorin 1980*a*). The principal *raison d'être* for our idealizations is simplicity, for in the first practical application of a new method the simplest example is most appropriate. Simplicity, moreover, provides for economy in computations. The computational techniques developed on this basis should be of benefit to future work on more involved problems.

Notation

A	area
c	$(T-T_u)/(T_b-T_u)$, reaction progress parameter or reactedness
d_j	influence factor of a vortex sheet
f	V_b/V_c, fractional volume of burned medium in a cell
F	$d\zeta/dZ$, differential transformation function

h length of a vortex sheet
h_c side length of a cell
G Green function
H width of the channel
k time step
L reference length
i $(-1)^{\frac{1}{2}}$
J_b number of vortex blobs
J_s number of source blobs
$\boldsymbol{n}$ unit vector normal to solid boundaries
p $p'v'/\boldsymbol{u}'^2_\infty$, non-dimensional pressure
p' pressure
$\boldsymbol{r}$ (x, y), position vector
r_0 blob core radius
R $\boldsymbol{u}'_\infty H/2v'\mu$, Reynolds number
$\boldsymbol{s}$ unit vector tangential to solid walls
S normal burning velocity
t' time
t $t'\boldsymbol{u}'_\infty/L$, non-dimensional time
T temperature
$\boldsymbol{u}$ (u, v) non-dimensional velocity vector
u non-dimensional velocity component in the x-direction
$\boldsymbol{u}'_\infty$ inlet velocity
v non-dimensional velocity component in the y-direction
v' specific volume
V volume
W $u - \mathrm{i}v$, complex velocity
x, y non-dimensional Cartesian space coordinates
Z $x + \mathrm{i}y$, complex position coordinate in physical plane

γ $\int \xi \,\mathrm{d}y$, circulation per unit length
Γ $\int \xi \,\mathrm{d}A$, circulation
δ Dirac δ-function
δ_s thickness of the numerical shear layer
Δ source strength
ϵ local rate of expansion
ζ complex position coordinate in the transformed plane
η Gaussian random variable
μ dynamic viscosity
ν T_b/T_u, temperature, or specific volume ratio
ξ $\nabla \times \boldsymbol{u}$, vorticity
σ $(2k/R)^{\frac{1}{2}}$, standard deviation
ϕ velocity potential
ψ stream function
∇^2 Laplacian operator

Subscripts

b burnt medium
c cell
f flame front
i point in space
j vortex element
p potential flow produced by $\boldsymbol{u}_\infty$
s source velocity
u unburnt medium
ϵ produced by combustion
ξ produced by turbulence
σ due to combustion

Superscript

* complex conjugate

1. Problem

According to the arguments presented in the Introduction, the problem we treat is formulated on the basis of the following idealizations:

(i) the flow is two-dimensional, i.e. strictly planar;

(ii) the flowing substance consists only of two incompressible media; the unburned mixture and the burned gas;

(iii) the flame is treated as a constant-pressure deflagration acting as an interface between the two media, and propagating locally at a prescribed normal burning velocity;

(iv) the exothermicity of combustion is manifested entirely by an increase in specific volume associated with the transformation of one component medium into the other.

Thus the following physical phenomena are, with respect to each idealization, completely neglected:

(i) three-dimensional effects, in particular vortex stretching;

(ii) compressibility effects, in particular acoustic wave interactions;

(iii) chemical kinetic effects and molecular transport processes, governing the structure of the flame and its propagation velocity, as well as such secondary effects as the generation of vorticity due to the interaction between the density jump across the flame front and the pressure gradient in the flow field;

(iv) thermal effects, in particular all the thermodynamic properties of the substance and the heat transfer processes.

It should be noted that the model described is consistent with the well known model of thin-flame, or infinitely fast kinetics, used widely for the analysis of mixed controlled turbulent combustion.

As a consequence of these idealizations, the continuity and the Navier–Stokes equations governing the flow field can be expressed in the following simple form:

$$\nabla \cdot \boldsymbol{u} = \epsilon(\boldsymbol{r}_f), \tag{1.1}$$

$$D\boldsymbol{u}/Dt = R^{-1}\nabla^2\boldsymbol{u} - \nabla p, \tag{1.2}$$

where $\boldsymbol{u} = (u, v)$ is the velocity vector normalized by the inlet velocity $\boldsymbol{u}'_\infty$, ϵ is the corresponding local rate of expansion, $\boldsymbol{r} = (x, y)$ is the position vector normalized by L, the reference length, t is the time normalized by $L/\boldsymbol{u}'_\infty$, R is the Reynolds number, p is the pressure normalized by $\boldsymbol{u}'_\infty/v'$, v' denoting the initial specific volume of the medium, and subscript f refers to the flame front, while

$$D/Dt = \partial/\partial t + \boldsymbol{u}\cdot\nabla$$

is the substantial derivative, ∇^2 is the Laplacian, and ∇ is the usual differential vector operator.

The flow field is specified by the solution of these equations, subject to the boundary conditions

$$\boldsymbol{u} = 0 \quad \text{along all solid boundaries,} \tag{1.3}$$

$$\boldsymbol{u} = (1, 0) \quad \text{at the inlet.} \tag{1.4}$$

The distribution of ϵ is determined by the location of the flame front, $\boldsymbol{r}_f$, which is governed by the flame propagation equation

$$\partial\boldsymbol{r}_f/\partial t = S_u\boldsymbol{n}_f + \boldsymbol{u}, \tag{1.5}$$

where S_u is the normal burning velocity, and $\boldsymbol{n}_f$ is the unit vector normal to the flame surface.

2. Procedure

The procedure is based on the principle of fractional steps (see, for example, Lie & Engel 1880; Samarski 1962), according to which the governing equations are split into a sum of elementary components, and the solution is determined by treating these components in succession.

The essential element used for this purpose is the vorticity. .

$$\xi = \nabla \times \boldsymbol{u}, \tag{2.1}$$

which is introduced by expressing equation (1.2) in terms of its curl, the vortex transport equation,

$$D\xi/Dt = R^{-1}\nabla^2\xi, \tag{2.2}$$

since $\nabla \times \nabla p \equiv 0$.

The flow is thus described by equations (1.1), (2.1) and (2.2).

Equations (1.1) and (2.1) are used to determine the velocity field, $\boldsymbol{u}(\boldsymbol{r})$, while, in accordance with the principal feature of the r.v.m., equation (2.2) is used to update the vorticity field, $\xi(x, y)$. Then $\epsilon(x, y)$ is determined by the flame propagation algorithm we developed for the solution of equation (1.5).

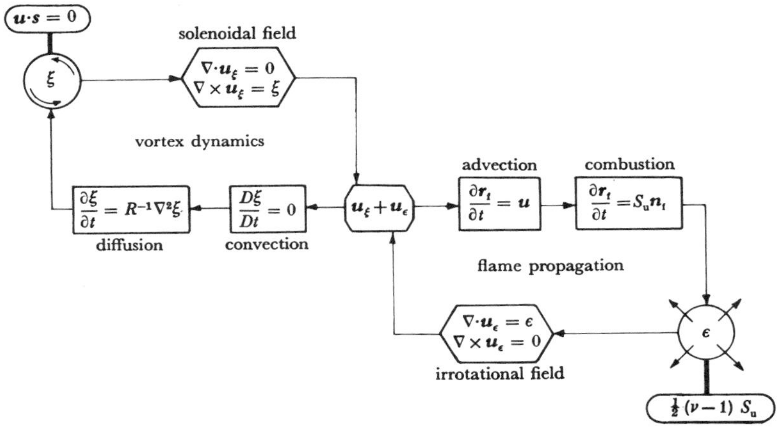

FIGURE 2. Structure of the algorithm.

Thus $\boldsymbol{u}$ is decomposed into a divergence-free vector field $\boldsymbol{u}_\xi$ and a curl-free field $\boldsymbol{u}_\epsilon$, where

$$\boldsymbol{u} = \boldsymbol{u}_\xi + \boldsymbol{u}_\epsilon. \tag{2.3}$$

In doing this, we exploit the Hodge decomposition theorem (see Batchelor 1967; Chorin & Marsden 1979). The governing equations for $\boldsymbol{u}_\xi$ and $\boldsymbol{u}_\epsilon$ are then obtained immediately by the substitution of equation (2.3) in equation (1.1) and (2.1):

$$\nabla\cdot\boldsymbol{u}_\xi = 0, \quad \nabla\times\boldsymbol{u}_\xi = \xi, \tag{2.4, 2.5}$$

and

$$\nabla\cdot\boldsymbol{u}_\epsilon = \epsilon, \quad \nabla\times\boldsymbol{u}_\epsilon = 0. \tag{2.6, 2.7}$$

Both $\boldsymbol{u}_\xi$ and $\boldsymbol{u}_\epsilon$ are required to satisfy independently the zero normal velocity boundary condition, namely

$$\boldsymbol{u}_\xi\cdot\boldsymbol{n} = 0, \quad \boldsymbol{u}_\epsilon\cdot\boldsymbol{n} = 0, \tag{2.8, 2.9}$$

where $\boldsymbol{n}$ is the unit vector normal to the walls.

However, only the total velocity, $\boldsymbol{u}$, is required to satisfy the no-slip condition

$$\boldsymbol{u}\cdot\boldsymbol{s} = 0, \tag{2.10}$$

where $\boldsymbol{s}$ is the unit vector tangential to the walls.

The structure of the algorithm is described schematically in the form of a flow diagram in figure 2. There are two loops, one for handling *vortex dynamics*, and the other for *flame propagation*. These are linked together to yield the total velocity field. Key elements in the first loop are vortices, generated by the no-slip boundary condition at the walls and transported by diffusion and convection, the fractional steps of equation (2.2). Key elements in the second loop are sources of specific volume, generated by the flame propagation process and monitored by combustion, the fractional steps of equation (1.5).

3. Vortex dynamics

The mechanism of turbulence is described in essence by vortex dynamics. This process is evaluated here by first determining the velocity field $\boldsymbol{u}$, produced by a given vorticity distribution, $\xi(x, y)$, according to equations (2.4) and (2.5), with the zero normal velocity boundary condition, equation (2.8), and then updating the vorticity field in accordance with the vortex transport equation, equation (2.2), implementing at the same time the non-slip boundary condition, equation (2.10).

As the principal feature of the r.v.m., the flow field is expressed for this purpose in terms of discrete elements, the so-called vortex blobs and vortex sheets. Their properties are presented here in turn.

(a) Vortex blobs

To derive the properties of vortex blobs, equation (2.5) is expressed in terms of the stream function, ψ,

$$\nabla^2\psi = -\xi, \tag{3.1}$$

where

$$u = \partial\psi/\partial y; \quad v = -\partial\psi/\partial x, \tag{3.2}$$

so that equation (2.4) is satisfied exactly.

The vorticity is then described in terms of

$$\xi_j = \Gamma_j\,\delta(\boldsymbol{r}-\boldsymbol{r}_j), \tag{3.3}$$

where δ is the Dirac δ-function and

$$\Gamma_j = \lim_{\Delta A_j \to 0} \int_{\Delta A_j} \xi_j \, \mathrm{d}A \tag{3.4}$$

is the circulation of a vortex at $\boldsymbol{r}_j$, while ξ_j is acting on area ΔA_j. The solution of equation (3.1) is given by the Green function

$$G(\boldsymbol{r}, \boldsymbol{r}_j) = (\Gamma_j/2\pi)\ln|\boldsymbol{r}-\boldsymbol{r}_j|, \tag{3.5}$$

representing the field of a potential vortex.

Equation (3.5) can then be used to construct a solution to equation (3.1) for a general distribution of ξ in the form

$$\psi(x, y) = \int_A G(\boldsymbol{r}-\boldsymbol{r}_j)\,\xi(\boldsymbol{r}_j)\,\mathrm{d}A, \tag{3.6}$$

where A is the area of the flow field. The above integral can be evaluated as a sum of all the contributions of ξ, after it had been partitioned into discrete elements ξ_j. The elementary vorticity, ξ_j, is a function of small support that tends to a δ-function as the area where it exists, ΔA_j, approaches zero. This process requires smoothing of the function in equation (3.5) to eliminate the singularity at its centre (Chorin 1973). Thus the integral in equation (3.6) becomes

$$\psi = \sum_j G_j \Gamma_j, \tag{3.7}$$

in which case

$$\Gamma_j = \int_{\Delta A_j} \xi_j \, dA,$$

where ΔA_j is finite, while G_j is the corresponding smooth Green function at r_j (Chorin 1973; Hald 1979).

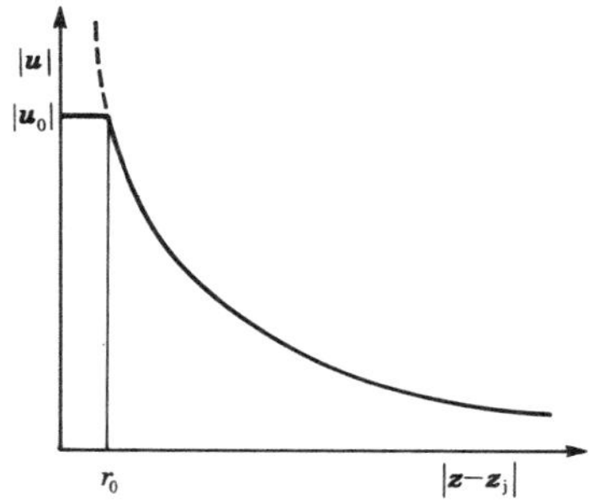

FIGURE 3. Velocity distribution of a blob.

The elementary component of the flow field specified by equation (3.7) is called a vortex blob. The velocity field it produces in a free space, i.e. one without boundaries, is obtained by substituting equation (3.5) into equation (3.2) and smoothing G around the centre. For this purpose the velocity vector is expressed, in terms of complex variables as follows:

$$W_\psi(Z, Z_j) = \frac{-\mathrm{i}\Gamma_j |Z-Z_j|}{2\pi \max(|Z-Z_j|, r_0)} \frac{1}{(Z-Z_j)}, \tag{3.8}$$

where $W = u - \mathrm{i}v$, $\mathrm{i} = (-1)^{\frac{1}{2}}$, $Z = x + \mathrm{i}y$, while r_0 is the cut-off radius, i.e. the radius of the core within which $|\boldsymbol{u}|$ is constant, in compliance with the smoothing requirement for the function expressed by equation (3.5). The velocity distribution of a blob is displayed in figure 3.

To satisfy the boundary condition given by equation (2.8), we use conformal mapping to transform the flow field into the upper-half ζ-plane, and add the velocity produced by the image of vortex ζ_j. The corresponding velocity field in the ζ-plane produced by a vortex blob at ζ_j is thus given by

$$W_\xi(\zeta, \zeta_j) = W_\psi(\zeta, \zeta_j) - W_\psi(\zeta, \zeta_j^*), \tag{3.9}$$

where $W_\psi(\zeta, \zeta_j)$ is given by equation (3.8), and an asterisk denotes a complex conjugate.

The boundary condition of equation (1.4) is satisfied by the velocity $W_\mathrm{p}(\zeta)$ of the potential flow produced by a unit velocity at the inlet. The total velocity produced by a set of J_b vortex blobs, including the effect of flow at the inlet, is thus

$$W_\xi(\zeta) = W_\mathrm{p}(\zeta) + \sum_{j=1}^{J_\mathrm{b}} W_\xi(\zeta, \zeta_j). \tag{3.10}$$

Then, to deduce the solution in the physical domain, the Z-plane, one applies the Schwarz–Christoffel theorem to specify the differential of the transform function,

$$\mathrm{d}\zeta/\mathrm{d}Z = F(\zeta), \tag{3.11}$$

for a given geometry of the flow field. Since

$$W(Z) = W(\zeta)\,F(\zeta), \tag{3.12}$$

the velocity vector, $\boldsymbol{u}_\xi$, in equation (2.3) is thus determined.

The vorticity, $\zeta(x, y)$, is updated at every computational time step, k, by solving equation (2.2) in fractional steps made up of the contribution of the convection operator

$$D\xi/Dt = 0 \tag{3.13}$$

and that of the diffusion operator

$$\partial\xi/\partial t = R^{-1}\nabla^2\xi. \tag{3.14}$$

According to equation (3.13), vortex blobs move at an appropriate particle velocity specified by equation (2.3).

The solution corresponding to a time step, k, of a one-dimensional component of the diffusion equation (3.14), when the initial condition is given by the Dirac delta function, $\delta(0)$, is the Green function

$$G(x, k) = (4\pi k/R)^{-\frac{1}{2}} \exp\,(-Rx^2/4k). \tag{3.15}$$

This is the probability density function of a Gaussian random variable with zero mean and a standard deviation of $\sigma = (2k/R)^{\frac{1}{2}}$! Thus, if the initial vorticity is split into a set of discrete vortex elements and each is given a displacement from the origin by an amount drawn from a set of Gaussian random numbers of an appropriate variance, it provides an approximation to equation (3.15) by sampling. When a general distribution of vorticity $\xi(x)$ is given, the exact solution of equation (3.14), after a time interval k, is

$$\gamma(x) = \int_A G(x-x', k)\;\xi(x')\;\mathrm{d}x' \tag{3.16}$$

where γ denotes the circulation per unit length, while G is given by equation (3.15). The probabilistic counterpart of this solution is obtained by displacing each vortex element from its position x' through a distance η_i. The random walk is then constructed by repeating this procedure at each time step. Two-dimensional random walk is treated in essentially the same way, the vortex elements being moved in two mutually perpendicular directions x and y, by two independent Gaussian random variables with zero mean and a standard deviation of $\sigma = (2k/R)^{\frac{1}{2}}$.

The convection and diffusion contributions in the Z-plane are combined, according to equation (2.2), by the summation

$$Z_j(t+k) = Z_j(t) + W^*(Z_j)\;k + \eta_j, \tag{3.17}$$

where $W = W_\xi + W_\epsilon$ and $\eta_j = \eta_x + \mathrm{i}\eta_y$ or, in the ζ-plane, by using its transform

$$\zeta_j(t+k) = \zeta_j(t) + W^*(\zeta_j)\;F^*(\zeta_j)\;F(\zeta_j)\;k + \eta_j F(\zeta_j). \tag{3.18}$$

Since the velocity is calculated in the ζ-plane by implementing equation (3.10), the use of equation (3.18) is more straightforward and hence more economical than that of equation (3.17).

To satisfy the no-slip boundary condition expressed by equation (1.3), the velocity, W, has to be calculated at points along the wall. The points are selected to be a distance h apart along each wall. Wherever the tangential velocity u at wall is not zero, a vortex with a circulation $u_w h$ is created and included in the computations at the next time step, according to equation (3.17) or equation (3.18). However, this procedure of vorticity creation is not accurate since on the average one-half of the newly created blobs are lost through diffusion across the wall. This implies that Kelvin's theorem is not satisfied exactly and the accuracy near the wall is poor. Furthermore, vortex blobs do not provide a good description of the flow near solid walls where velocity gradients are very high, because inside the core of a blob the velocity is considered to be constant. This motivates the introduction of vortex sheets to take up the role of blobs in shear layers at the walls.

(b) Vortex sheets

If we take x to be the direction along a wall and y the normal to it, the following two conditions are known to prevail in the shear layer immediately adjacent to the wall:

(i)
$$\partial v/\partial x \ll \partial u/\partial y; \tag{3.19}$$

(ii) diffusion in the x-direction is negligibly small in comparison to convection in this direction.

A vortex element constructed on the basis of these conditions is referred to as the vortex sheet.

As a consequence of equation (3.19), equation (2.5) is reduced to

$$\xi = -\partial u/\partial y. \tag{3.20}$$

The foregoing, in conjunction with equation (2.4), determines $\boldsymbol{u}_\xi(\boldsymbol{r})$ as follows. Integrating equation (3.20) from $y = \delta_s$, the outer edge of the numerical shear layer at the wall, to y_i, one obtains,

$$u_\delta(x_i) - u(x_i, y_i) = -\int_{y_i}^{\delta_s} \xi \, dy, \tag{3.21}$$

where $u_\delta = u$ at $y = \delta_s$.

The integral in (3.21) can be transformed into a summation by partitioning the value of ξ along y and defining the circulation per unit length of a vortex sheet as

$$\gamma_j = \lim_{\Delta y \to 0} \int_{y_i}^{y_i = \Delta y} \xi dy. \tag{3.22}$$

If a sheet has a length h, then its circulation, Γ_j, is

$$\Gamma_j = h\gamma_j, \tag{3.23}$$

and from equations (3.21) and (3.22) the velocity jump across it, Δu_j, per unit sheet length is

$$\Delta u_j = \gamma_j. \tag{3.24}$$

Unlike the 'elliptic' flow modelled by vortex blobs, where the effect of each blob extends throughout the field, the zone of influence of a vortex sheet is, as a consequence of equation (3.21), restricted to the 'shadow' below it. This zone is indicated by the regions with vertical hatching in figure 4. Thus, the flow velocity at a point (x_i, y_i), where $y_i < y_j$, is determined by the relation

$$u(x_i, y_i) = u_\delta(x_i) - \sum_j \gamma_j d_j, \tag{3.25}$$

a summation counterpart of equation (3.21) according to equations (3.22) and (3.24), while

$$d_j = 1 - |x_i - x_j|/h$$

is the influence factor of sheet j on point i, expressing the fraction of the length of the sheet extending over the zone of dependence over point i, indicated by horizontal hatching in figure 4.

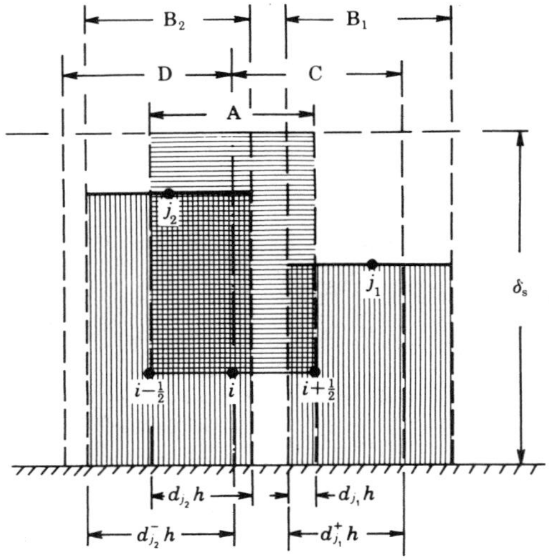

FIGURE 4. Geometry of interdependence in the numerical shear layer: A, zone of dependence over point i; B, zone of influence under sheet j; C, zone of dependence around point $i+\frac{1}{2}$; D, zone of dependence around point $i-\frac{1}{2}$.

The value of v is determined by the integration of the expression

$$v = -\frac{\partial}{\partial x}\int_0^{y_i} u\,dy, \tag{3.26}$$

obtained from equation (2.4) by using $u(x_i, y_j)$ as evaluated from equation (3.24).

For this purpose we introduce

$$I = \int_0^{y_i} u\,dy = u(x_i)\,y_i - \int_0^{y_i} y\,du = u(x_i)\,y_i - \sum_j \gamma_j d_j y_j, \tag{3.27}$$

where, by taking advantage of equation (3.24), Δu has been replaced by $\gamma_j d_j$. In finite-difference form, equation (3.26) then becomes

$$v(x_i, y_i) = -(I^+ - I^-)/h, \tag{3.28}$$

where, according to equation (3.27), by using equation (3.25) for $u(X_i)$,

$$I^{\pm} = u_\delta(x_i \pm \tfrac{1}{2}h)\,y_i - \sum_j y_j^0 \gamma_j d_j^{\pm}, \tag{3.29}$$

while, as indicated in figure 4,

$$d_j^{\pm} = 1 - (x_i \pm \tfrac{1}{2}h - x_j)/h$$

and

$$y^0 = \min(y_i, y_j).$$

The motion of the sheets is governed by an equation identical to (3.17), but with $\boldsymbol{u}$ evaluated from equations (3.25) and (3.29) while $\eta_i = 0+i\eta_y$, in accordance with condition (ii) stated at the beginning of this section. To make sure that the motion of a vortex sheet is matched with the vortex blob into which it transforms, a correction term of $-\frac{1}{2}\gamma_j$ has to be added to the expression for $u(x_i, y_i)$ given by equation (3.25) to account for the effect of the image of the blob. The paper introducing the vortex sheet method (Chorin 1978) contains information on techniques to reduce the statistical error and speed up the convergence of the algorithm.

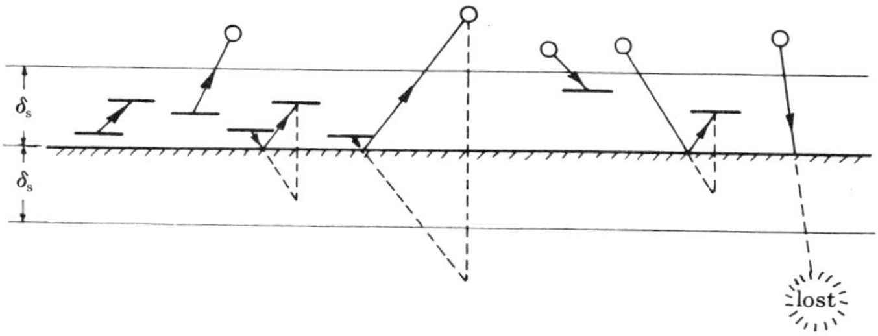

FIGURE 5. Transformations of vortex elements in and around a numerical shear layer at the wall.

(c) *Algorithm*

The foregoing concepts are implemented as follows.

First the value of h, the sheet length specifying the spatial resolution, is chosen. The value of the time step, k, is then fixed in accordance with the Courant stability condition, $k \leqslant h/\max|\boldsymbol{u}|$ (Chorin 1980*a*). For a given Reynolds number, this specifies the standard deviation σ. The thickness of the numerical shear layer δ_s is then taken as a multiple of σ whereby, as shown in figure 5, the loss of vortex blobs due to their random walk is minimized. Finally, the number of sheets initially in the stack is chosen, limiting the maximum allowable value for γ.

At time zero only the incoming flow $u_p(\zeta)$ exists. The resulting velocity along the wall is fixed by the potential flow solution of equation (3.1). The displacement of the sheets in the numerical shear layer is then calculated, by using equation (3.17) with velocity specified by equations (3.25) and (3.28). The various possibilities that may occur owing to vortex sheet displacement are illustrated in figure 5. When a sheet gets out of the boundary layer, it becomes a blob with a total circulation adjusted according to equation (3.23).

The core radius, r_0, is then fixed in such a way that the no-slip boundary condition is satisfied. To do so with a minimum error, one sets $r_0 > \delta_s$. The velocity at the wall produced by the blob and its image is then, in accordance with equation (3.9),

$$u_0 = \Gamma_j/\pi r_0,$$

whence, by virtue of equations (3.23) and (3.24) with $\Delta u_j = u_0$,

$$r_0 = h/\pi, \tag{3.30}$$

providing an explicit relation between the length of the vortex sheet and the core radius of a vortex blob.

If a sheet gets out on the other side of the wall, it becomes restored by its mirror image either in the shear layer as a sheet or in the flow field as a blob, as depicted in figure 5.

Corresponding displacements of vortex blobs are calculated by the use of equation (3.18) with their velocities evaluated from equation (3.10). Again figure 5 displays the various ways in which a blob can be transformed into a sheet. The last possibility of losing a vortex blob is minimized by the right choice of δ_s, as already pointed out.

Once the position and strength of both the sheets and blobs are established, the flow field at a given time step is fully determined. It should be noted that vortex blobs appear only as a consequence of the displacement of vortex sheets outside the boundary layer, modelling the mechanism of the generation of turbulence under actual flow conditions.

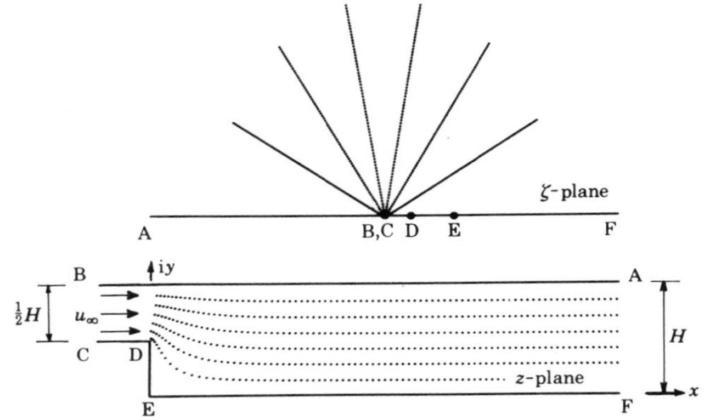

FIGURE 6. Streamlines pattern of initial flow in transformed plane, and physical plane of a channel with a step expansion.

(d) *Results*

The Z-plane and ζ-plane for flow over a rearward-facing step are present in figure 6. The functions W_p and F are

$$W_p(\zeta) = H/\pi\zeta \tag{3.31}$$

and

$$F(\zeta) = \frac{\pi\zeta}{H}\left(\frac{\zeta-4}{\zeta-1}\right)^{\frac{1}{2}}, \tag{3.32}$$

where, as shown in figure 6, H is the height of the channel.

The results of computations for turbulent flow behind a step of the same geometrical proportions as figure 1, corresponding to a Reynolds number of 10^4 at inlet, or for a channel 2.54 cm wide in the experimental apparatus, $u'_\infty = 6 \text{ m s}^{-1}$, are shown in figure 7. Included are two sequential series of computer outputs. Series (*a*) shows the development of the flow field by presenting vortex velocity vector fields tracing the motion of all the vortex blobs included in the solution at successive time intervals, each equal to 50 computational steps of $0.1(H/2u'_\infty)$ s. Series (*b*) shows the growth of a large-scale eddy traced at time intervals equal to five computational steps.

A velocity vector is represented in figure 7 by a line segment providing information on its magnitude and direction. However, instead of being furnished with the conventional arrowhead, it is attached at its origin to a small circle denoting the location of the vortex blob to which it pertains.

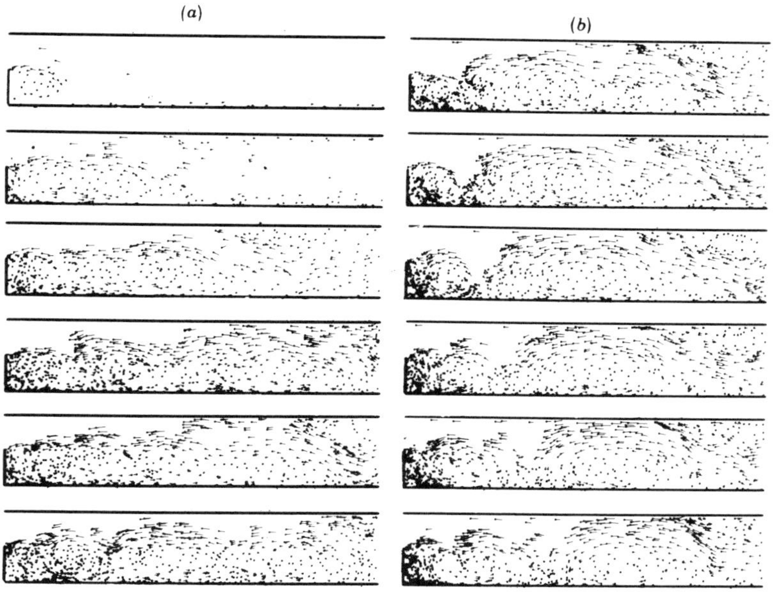

FIGURE 7. Sequential series of computer plots displaying vortex velocity fields in turbulent flow behind a step at inlet Reynolds number $R = 10^4$: (a) development of the flow field; (b) growth of a large-scale eddy.

4. Flame propagation

According to idealization (iii), the flame front is treated as an interface across which reactants are transformed into products at a rate controlled by the normal burning velocity. The method used for tracing the motion of such an interface was developed by Chorin (1980*b*) and implemented with the help of the algorithm of Noh & Woodard (1976).

The flow field is divided for this purpose into square cells by a grid of mesh size h_c. The fraction of volume, V, occupied in a given cell by the burned medium is expressed in terms of a number

$$f \equiv V_b/V_c, \tag{4.1}$$

where subscripts b and c refer, respectively, to the burnt medium and the cell. In terms of specific volumes, v_i' (i = b, c, u, and the last refers to the unburnt medium),

$$f = (1 - v_u'/v_c')/(1 - v_u'/v_b'). \tag{4.2}$$

Since, because of idealization (iii), the flame is treated as a constant-pressure deflagration, f can be expressed in terms of the usual reaction progress parameter

$$c \equiv \frac{T_c - T_u}{T_b - T_u} = \frac{1}{\nu - 1}\left(\frac{T_c}{T_u} - 1\right), \tag{4.3}$$

where

$$\nu \equiv T_b/T_u = v_b'/v_u', \tag{4.4}$$

while T can be considered to represent either absolute temperature (if the change in molecular mass is negligible) or temperature divided by the molecular mass.

Thus, with the use of the perfect gas equation of state, equations (4.2), (4.3) and (4.4) yield

$$c = f/\{\nu + (\nu - 1) f\}, \tag{4.5}$$

specifying, in effect, the temperature distribution, for, as a consequence of idealization (ii), ν is a constant for a given combustible medium.

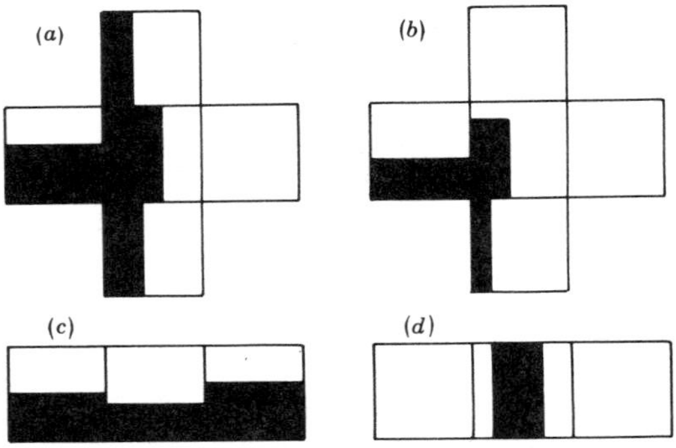

FIGURE 8. Elementary components of an interface recognized by the algorithm.

Thus, $f = 0$, 1 means that there is, respectively, either unburnt or burnt medium in the cell, while fractional values of f indicate cells containing the interface. The particular geometry of the interface is deduced, depending on the f-numbers of neighbouring cells. In this connection, as illustrated in figure 8, proper provisions are included in the algorithm for four possibilities:

(*a*) vertical interface;
(*b*) horizontal interface;
(*c*) rectangular corner;
(*d*) neck.

As a consequence, the interface is made up of horizontal and vertical line segments, yielding higher spatial resolution than h_c, the mesh size of the grid.

The motion of the interface, or flame propagation, is described by equation (1.5). By virtue of the principle of fractional steps, its effects are split into two components:

(*a*) advection, prescribed by

$$\partial \boldsymbol{r}_f/\partial t = \boldsymbol{u}; \tag{4.6}$$

(*b*) combustion, prescribed by

$$\partial \boldsymbol{r}_f/\partial t = S_u \boldsymbol{n}_f, \tag{4.7}$$

providing a proper set-up for the inclusion of the effects of

(*c*) exothermicity.

Algorithms for each of these processes are presented here in sequence.

(*a*) *Advection*

The advection step is the passive displacement due to the velocity field. It is evaluated by calculating first the velocity components at mid-points on the sides of the cell, as shown in figure 9. The interface is then transported in two fractional steps, one horizontal and one vertical, changing the f-number of the cell by an amount proportional to corresponding displacements in the time step, k_c. The algorithm is stable whenever the Courant condition, $k_c \leqslant h_c/\max|\boldsymbol{u}|$, is satisfied (see Noh & Woodward 1976).

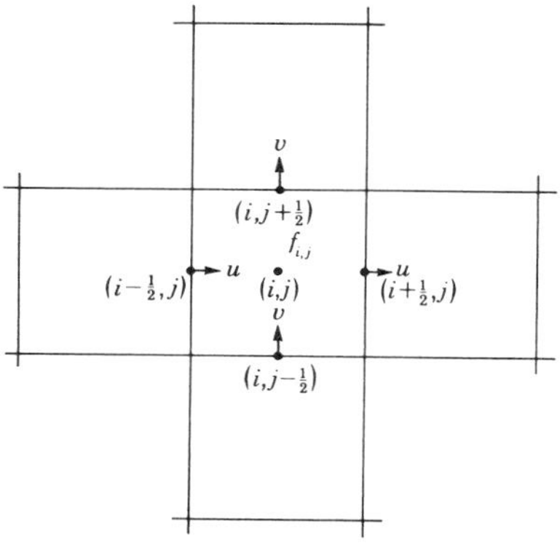

FIGURE 9. Velocity components used in the advection algorithm to determine the motion of the interface in cell (i, j).

(b) *Combustion*

The combustion step is the advancement of the front due to consumption of the unburned medium. The front moves in the direction of its normal with a relative velocity taken here as a constant, equal to the appropriate laminar burning velocity of the mixture, S_u. The corresponding motion of the interface is evaluated by the implementation of the Huygens principle, with the use of the advection algorithm. At the nth computational step one calculates for this purpose the displacements due to S_u in eight directions – the four sides and four corners of each cell – so that all of the cell's neighbours are affected. For a given cell at (i, j) this results in eight new f-numbers. The value assigned to it is then

$$f_{ij}^{n+1} = \max_{0 \leqslant l \leqslant 8} f_{ij}^{(l)}, \tag{4.8}$$

where $l = 1, \ldots, 8$, while $f_{ij}^{(0)} = f_{ij}^{n}$. It should be noted that the algorithm provides, in effect, information on the displacement of the interface due to its motion at a given velocity normal to its frontal surface, without having to determine its actual direction (see Chorin 1980*b*).

(c) *Exothermicity*

Mechanical effects of the exothermic process are manifested by volumetric expansion behind the flame front. The velocity field induced thereby is governed by equations (2.6), (2.7) and (2.9).

As for vortex blobs, equations (2.6) and (2.7) are solved by superposition. A velocity potential, ϕ, is introduced for this purpose, so that

$$u = \partial\phi/\partial x, \quad v = \partial\phi/\partial y, \tag{4.9}$$

satisfying exactly equation (2.7). The governing equation for ϕ,

$$\nabla^2\phi = \epsilon, \tag{4.10}$$

is then obtained immediately by substitution of equation (4.8) into equation (2.6).

The solution of this equation is given by

$$\phi(\boldsymbol{r}) = \int_A G(\boldsymbol{r}, \boldsymbol{r}')\, \epsilon(\boldsymbol{r}')\, dA, \tag{4.11}$$

where
$$G(\boldsymbol{r}, \boldsymbol{r}') = (1/2\pi) \ln |\boldsymbol{r} - \boldsymbol{r}'|$$
is the Green function.

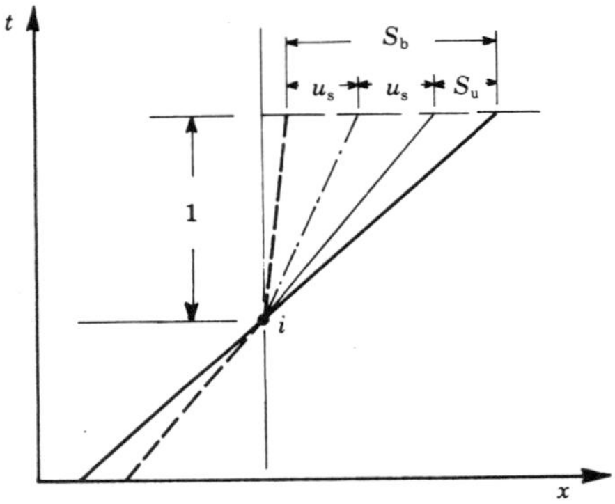

FIGURE 10. Kinematics of the flame front: ——, tangent to the flame front at point i; – – –, tangents to particle paths at point i.

By following the procedure used for integrating equation (3.6), the solution of equation (4.11) is approximated by the summation

$$\phi = \sum_j G(\boldsymbol{r}, \boldsymbol{r}_j)\, \Delta_j, \tag{4.12}$$

where Δ_j is the source strength,
$$\Delta_j = \int_{\Delta_j} \epsilon_j \, dA, \tag{4.13}$$

i.e. the rate of volumetric expansion induced by the source, while ϵ_j is the Dirac δ-function. As before, for the summation to converge, the Green function is smoothed around $\boldsymbol{r}_j$.

The concept of a source blob is conceived in analogy to a vortex blob. The velocity field produced by a source blob is, in effect, the same as that in figure 3. Thus the vel ocity produced in a free space by a source blob at Z_j is given by

$$W_\phi(Z, Z_j) = \frac{\Delta_j\, |Z - Z_j|}{2\pi \max(|Z - Z_j|, r_0)} \frac{1}{(Z - Z_j)}. \tag{4.14}$$

The boundary condition expressed by equation (2.9) is satisfied by adding the velocity produced by the image of each source on the other side of the wall. Thus the total velocity produced by a source blob in the ζ-plane is given by

$$W_\epsilon(\zeta, \zeta_j) = W_\phi(\zeta, \zeta_j) + W_\phi(\zeta, \zeta_j^*). \tag{4.15}$$

For J_s source blobs, the solution of equation (4.11) is approximated by the summation specified in equation (4.12), and the corresponding flow velocity is then

$$W_\epsilon(\zeta) = \sum_{j=1}^{J_s} W_\epsilon(\zeta, \zeta_j). \tag{4.16}$$

The strength of the source is adjusted to provide for the volumetric expansion specified by Δ_j (see equation (4.13)), preserving mass. In one-dimensional flow, as depicted in figure 10,

$$u_s = \tfrac{1}{2}(S_b - S_u), \tag{4.17}$$

where S_b is the flame speed as seen from the burned side and S_u is the flame speed as seen from the unburned side, whence, as a consequence of the continuity requirement,

$$S_b = \nu S_u, \tag{4.18}$$

where $\nu \equiv v'_b/v'_u,$

and one has $u_s = \tfrac{1}{2}S_u(\nu - 1).$ (4.19)

The source strength is then $\Delta_j = u_s h_c = \tfrac{1}{2}S_u h_c(\nu - 1),$ (4.20)

while the volumetric rate of combustion is given by

$$dV_\sigma/dt = h_c S_u = h_c^2 df_\sigma/dt, \tag{4.21}$$

where df_σ/dt is the rate of change in f due solely to combustion. This yields

$$\Delta_j = \tfrac{1}{2}h_c^2(\nu - 1)\, df_\sigma/dt. \tag{4.22}$$

At the same time, with reference to figure 3, the source velocity is identified with the velocity of the core,

$$u_0 = \Delta_j/2\pi r_0, \tag{4.23}$$

so that, as a consequence of equation (4.19), one obtains

$$\Delta = 2\pi r_0 u_s = \pi r_0 S_u(\nu - 1). \tag{4.24}$$

Thus, by virtue of equation (4.23), it follows that

$$r_0 = (h_c^2/2\pi S_u k_c)\Delta f_\sigma, \tag{4.25}$$

where the time derivative of f has been expressed in terms of the change in f evaluated for a given time step, k_c, by the implementation of the Huygens principle.

By virtue of equation (2.3), volumetric sources affect the velocity field. In particular, they modify the values of $\boldsymbol{u}$. This in turn induces changes in the sheet velocities, as evident from equations (3.24) and (3.27), giving rise to new vortex blobs, etc. The whole algorithm is thus interrelated, as described schematically in figure 3.

(d) *Results*

With the use of the r.v.m. a solution was obtained for turbulent flow with combustion in the tunnel behind a step, modelling the process recorded photographically in figure 1. The exothermicity of the propane–air mixture, with equivalence ratio of 0.5, was for this purpose expressed in terms of the temperature (or specific volume) ratio $\nu = 4.25$, and the laminar burning velocity was taken as $S'_u = 12\ \mathrm{cm\ s^{-1}}$, while, as before, the velocity at inlet was $\boldsymbol{u}'_\infty = 6\ \mathrm{m\ s^{-1}}$ corresponding to $R = 10^4$.

An example of the results is presented in figure 11. Like figure 7, it consists of two sequential series of computer outputs, depicting the variation of vortex velocity vector fields and flame fronts. The flame contour has been delineated for this purpose as a line of demarcation between cells where $f = 0$ and those where $f > 0$. Series (*a*) depicts the process of ignition in the turbulent flow field of figure 7 in a cell located at point (1, 1), i.e., on the centre line of the tunnel at a distance from the step equal to the width of the inlet channel. The sequence in the right column

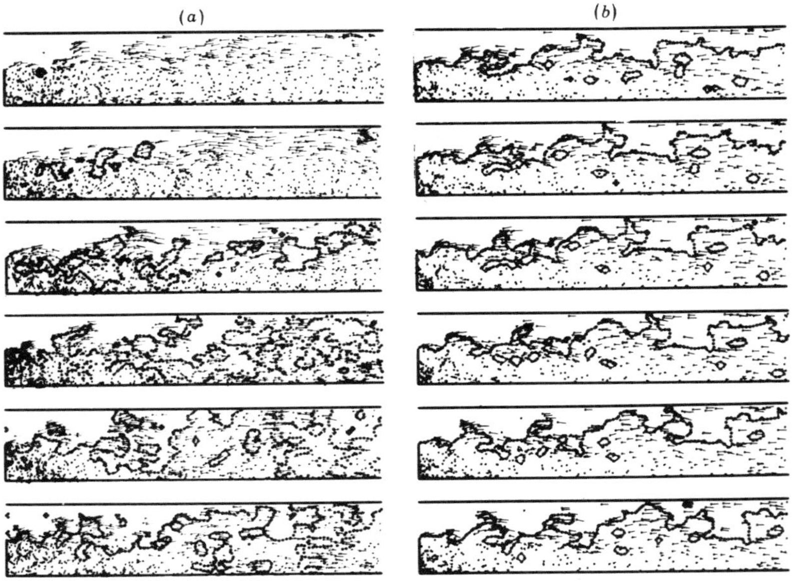

FIGURE 11. Sequential series of computer plots displaying vortex velocity fields and flame fronts in turbulent combustion behind a step at inlet Reynolds numbers $R = 10^4$ while $S_u = 0.02$ and $\nu = 4.25$, corresponding to a propane–air mixture at an equivalence ratio of 0.5; (*a*) ignition at point (1,1) in a fully developed turbulent flow; (*b*) 'steady state' turbulent flame propagation.

displays the 'steady flow condition' attained at time $t' = 26.102(H/2u'_\infty)$ s after ignition at the left bottom corner, point (0, 0), initiated at the moment when the medium was set in motion (hence the smaller number of vortex blobs). The number of computational time steps $0.05(H/2u'_\infty)$ s, between the solutions displayed here was 40 for series (*a*) and four for series (*b*).

In light of the stringent idealizations on which the computations are based, the agreement between the numerical model and the experimental observations is indeed remarkable. The r.v.m. is evidently capable of reproducing the essential features of the flow field associated with turbulent combustion as observed by schlieren photography, providing thereby a clarification of the essential mechanism of the process. At this stage one cannot expect, of course, more than a qualitative agreement. Quantitative modelling of stochastic turbulent flow parameters has to be left for future study.

5. Conclusions

The eminent suitability of r.v.m. for the study of the fluid mechanic properties of turbulent combustion has been demonstrated. The main advantage of the method is that it is unencumbered by numerical diffusion. Thus, all the instabilities in the flow field that arise as a characteristic feature of turbulence can be sustained without artificial damping, permitting their effects to be traced without undue distortion. Moreover, by using as the building block the mechanical properties of the essential ingredient of turbulence, the elementary eddy, the r.v.m. is capable of modelling the intrinsic physical properties of the flow system, subject only to restrictions introduced at the outset by the simplifying idealizations.

As a consequence, our analysis displays the following features of turbulent combustion:

(i) fluid mechanical processes of the formation of large-scale turbulent flow structure;

(ii) a rationale for the role played by the intrinsic instability of the flow system in stabilizing the flame, the basic mechanism of a blunt-body flame holder;

(iii) fluid mechanical processes of ignition in turbulent flow of pre-mixed gases;

(iv) detailed features of entrainment and mixing as principal means for the control of the combustion process;

(v) the mechanism of exothermic processes in turbulent combustion.

The authors wish to express their appreciation to the programme director of the research project under which these studies were in part conducted, Dr C. J. Marek of NASA Research Center, for his encouragement and support.

This work was supported by the Engineering, Mathematical, and Geoscience Division of the Department of Energy under contract W-7405-ENG-48, and by the National Aeronautics and Space Administration under NASA grant NSG-3227.

Appendix 1. Vortex motion in the transformed plane

Trajectories of the vortices in the transformed ζ-plane are required for the evaluation of the velocity $W(Z)$ with the use of equations (3.10) and (3.12). This can be obtained by a stepwise conformal mapping of trajectories in the Z-plane, defined by equation (3.17), with the use of the inverse of the transformation function $Z = Z(\zeta)$, the integral form of equation (3.11):

$$Z(\zeta) = \int \frac{d\zeta}{F(\zeta)}, \tag{A 1.1}$$

which, for the geometry of figure 6, is

$$Z = \frac{H}{\pi}\left(\ln\frac{1+q}{1-q} - \tfrac{1}{2}\ln\frac{2+q}{1-q}\right), \tag{A 1.1a}$$

where $$q = (4-\zeta)/(1-\zeta).$$

However, the inverse of the foregoing,

$$\zeta = \zeta(Z), \tag{A 1.2}$$

is awkward and time-consuming to evaluate. Hence one has to resort to numerical methods to integrate equation (3.11) directly to calculate corresponding displacements in the ζ-plane from those in the Z-plane.

This procedure can be reduced substantially if one uses equation (3.18) to trace these vortices in the ζ-plane directly, thus eliminating the use of the Z-plane except for the presentation of the results. This equation is obtained from equation (3.12) to write $W(Z)$ in terms of ζ as

$$W^*(Z) = \{W(\zeta)\, F(\zeta)\}^* = W^*(\zeta)\, F^*(\zeta). \tag{A 1.3}$$

If $\zeta(t)$ is defined to be the map of $Z(t)$, then

$$\zeta(t) = \zeta(Z(t)), \tag{A 1.4}$$

and one can write

$$\xi(t+k) - \zeta(t) = \zeta(Z(t+k)) - \zeta(Z(t)),$$

or, taking the first term of the Taylor series expansion,

$$\zeta(t+k) - \zeta(t) = \{Z(t+k) - Z(t)\}\, \mathrm{d}\zeta/\mathrm{d}Z. \tag{A 1.5}$$

If Z is a vortex centre, then equation (3.17) specifies the change in Z_j. Using equations (A 1.3) and (3.11) and rearranging, one obtains equation (3.18) as

$$\zeta_j(t+k) = \zeta_j(t) + \{W^*(\zeta_j)\, F^*(\zeta_j)\, k + \eta_j\}\, F(\zeta_j). \tag{A 1.6}$$

Equations (A 1.6) and (3.10) provide all the necessary information about the flow field. Interestingly, the effect of the Z-plane geometry on the motion in the ζ-plane is preserved in terms of F and F^* in equation (A 1.6).

Appendix 2. Derivation of equation (4.23)

The total volume created by a set of sources distributed on the surface of a flame should provide for extra volume on the side of the products owing to the expansion of reactants as they burn. If the fluid leaves a source with a velocity u_s normal to the surface of the flame, then

$$\frac{\mathrm{d}V}{\mathrm{d}t} = \int_f u_s \boldsymbol{n}_f \cdot \mathrm{d}\boldsymbol{A}_f, \tag{A 2.1}$$

where $\mathrm{d}V/\mathrm{d}t$ is the rate of volume increase due to the sources, and $\boldsymbol{A}_f$ is the area of the flame surface. For two-dimensional flow, $\mathrm{d}\boldsymbol{A}_f$ is equal to $\boldsymbol{n}_f \mathrm{d}L_f$, where L_f is the length of the flame front, and both V and $\boldsymbol{A}_f$ are measured per unit length normal to the plane of the flow. Since u_s is constant for homogeneous systems as indicated by equation (4.18), it follows that

$$\frac{\mathrm{d}V}{\mathrm{d}t} = u_s \int_f \boldsymbol{n}_f \cdot \boldsymbol{n}_f \mathrm{d}L_f = u_s L_f. \tag{A 2.2}$$

Flame propagation by combustion, the reason for volume expansion, is expressed by the left-hand side of equation (1.5):

$$\partial \boldsymbol{r}_f/\partial t = \boldsymbol{n}_f S_u. \tag{A 2.3}$$

When integrating the equation (A 2.1) for the flame surface A_f, the volumetric rate of combustion is specified as

$$\int_f \frac{\partial \boldsymbol{r}_f}{\partial t} \cdot \mathrm{d}\boldsymbol{A} = \int_f S_u \boldsymbol{n}_f \cdot \mathrm{d}\boldsymbol{A}. \tag{A 2.4}$$

The left side of equation (A 2.4) can be written as

$$\frac{\mathrm{d}V_b}{\mathrm{d}t} = \int_f \frac{\partial \boldsymbol{r}_f}{\partial t}\, \mathrm{d}\boldsymbol{A}_f = \sum_{i,j} h_c^2 \frac{\mathrm{d}f_\sigma}{\mathrm{d}t}, \tag{A 2.5}$$

where i and j cover the whole field of a planar flow. The integral on the right side of equation (A 2.4) is evaluated by assuming a constant S_u, yielding an expression similar to the right side of equation (A 2.2). Thus equation (A 2.4) becomes

$$\sum_{i,j} h_c^2 \frac{df_\sigma}{dt} = S_u L_f. \tag{A 2.6}$$

One can write

$$\frac{dV}{dt} = \sum_{i,j} \Delta_{ij}$$

and by eliminating L_f between equations (A 2.6) and (1.2) it follows that

$$\Delta = \frac{u_s}{S_u} h_c^2 \frac{df_\sigma}{dt}. \tag{A 2.7}$$

However, from equations (4.18) and (4.19), one has

$$u_s = \tfrac{1}{2} S_u (\nu - 1),$$

and equation (4.22) follows immediately. By using equation (A 2.7), one avoids calculating the flame length as required by equation (A 2.2). Instead, one uses the computations of the combustion step in the flame propagation algorithm, described before, to obtain the rate of change in the volume due to combustion, as indicated in equation (A 2.5).

Appendix 3. Conservation of circulation in a variable density field

The flame front, according to the model presented here, is a constant-pressure discontinuity across which a sudden change in density takes place. Of crucial importance to our modelling technique is the fact that vortex blobs conserve their circulation upon crossing this discontinuity. Provided here is a proof of this property, restricted to two-dimensional geometry.

In a two-dimensional potential flow, with variable density, the vortex transport equation (see, for example, equation (4.6) in Chorin & Marsden (1979)),

$$D(\xi/\rho)/Dt = 0, \tag{A 3.1}$$

expresses the variation of vorticity with density along a particle path. However, the variation of the circulation, given by

$$\frac{D\Gamma}{Dt} = \frac{D}{Dt}\int \xi \, dA = \frac{D}{Dt}\int (\xi/\rho)\, \rho \, dA, \tag{A 3.2}$$

can be calculated by reversing the integration and differentiation in the above expression. Thus

$$\frac{D\Gamma}{Dt} = \int \frac{D(\xi/\rho)}{Dt} \rho \, dA + \int (\xi/\rho) \frac{D(\rho \, dA)}{Dt}. \tag{A 3.3}$$

However, $\rho \, dA$ is constant along a particle path and taking equation (A 3.1) into account, we obtain

$$D\Gamma/Dt = 0. \tag{A 3.4}$$

Hence, since vortex blobs follow particle paths, their circulation in a variable-density field is invariant.

References

Ashurst, W. T. 1979 Numerical simulation of turbulent mixing layers via vortex dynamics. *Proc. 1st Symp. on Turbulent Shear Flows* (ed. by Durst *et al.*), pp. 402–413. Berlin: Springer-Verlag.

Ashurst, W. T. 1981 Vortex simulation of a Model Turbulent Combustor. *Proc. 7th Colloquium on Gas Dynamics of Explosions and Ractive Systems. Prog. Astronaut. Aeronaut.* **76**, 259–273.

Batchelor, G. K. 1967 *An introduction to fluid mechanics*. Cambridge University Press.

Chorin, A. J. 1973 Numerical studies of slightly viscous flow. *J. Fluid Mech.* **57**,785–796.

Chorin, A. J. 1978 Vortex sheet approximation of boundary layers. *J. comput. Phys.* **27**, 428–442.

Chorin, A. J. 1980*a* Vortex models and boundary layer instability. SIAM *Jl scient. Stat. Comput.* **1**, 1–24.

Chorin, A. J. 1980*b* Flame advection and propagation algorithms. *J. comput. Phys.* **35**, 1–11.

Chorin, A. J., Hughes, T. J. R., McCracken, M. F. & Marsden, J. E. 1978 Product formulas and numerical algorithms. *Communs pure appl. Math.* **31**, 205–256.

Chorin, A. J. & Marsden, J. E. 1979 *A mathematical introduction to fluid mechanics*. Berlin: Springer-Verlag.

Cheer, A. Y. 1979 A study of incompressible 2-D vortex flow past a circular cylinder. Lawrence Berkeley Laboratory, LBL-9950.

Ganji, A. R. & Sawyer, R. F. 1980 An experimental study of the flow field of a two-dimensional premixed turbulent flame. *AIAA Jl.* (In the press.)

Hald, O. H. 1979 Convergence of vortex methods for Euler's equations. II. *SIAM Jl numer. Anal.* **16**, 5, 726–755.

Lie, S. & Engel, F. 1880 *Theorie der Transformationsgruppen* (three vols). Leipzig: Teubner.

McCracken, M. & Peskin, C. 1980 Vortex methods for blood flow through heart valves. *J. comput. Phys.* **35**, 183–205.

McDonald, H. 1979 Combustion modeling in two and three dimensions – some numerical considerations. *Prog. Energy Combust. Sci.* **5**, 97–122.

Mellor, A. M. 1979 Turbulent-combustion interaction models for practical high intensity combustors. In *17th Symposium (International) on Combustion*, pp. 377–387. Pittsburgh: The Combustion Institute.

Noh, W. T. & Woodward, P. 1976 SLIC (Simple line interface calculation). *Proc. 5th Int. Conf. Numer. Math. Fluid Mechanics*, pp. 330–339. Berlin: Springer-Verlag.

Roshko, A. 1976 Structure of turbulent shear flows: a new look. *AIAA Jl* **14**, 10, 1349–1357.

Samarski, A. A. 1962 An efficient method for multi-dimensional problems in an arbitrary domain. *Zh. Vych. Mat. mat. Fiz.* **2**, 787–811 (transl. as *U.S.S.R. Comput. Math. & Math. Phys.* 1963, 894–896 (1964)).

Williams, F. A. 1974 A review of some theoretical considerations of turbulent flame structure. Specialists meeting on Analytical and Numerical Methods for Investigation of Flow Fields with Chemical Reaction, Especially Related to Combustion. *AGARD PEP* 43*rd Meetings*, vol. II, p. 1, Liège, Belgium, pp. 1–125.

Williams, F. A. & Libby, P. A. (ed.) 1980 *Turbulent reacting flows. Topics in applied physics* **4**. Berlin, Heidelberg, New York: Springer-Verlag.

The Evolution of a Turbulent Vortex*

Alexandre Joel Chorin

Department of Mathematics, University of California, Berkeley, CA 94720, USA

Abstract. We examine numerically the evolution of a perturbed vortex in a periodic box. The fluid is inviscid. We find that the vorticity blows up. The support of the L_2 norm of the vorticity converges to a set of Hausdorff dimension ~ 2.5. The distribution of the vorticity seems to converge to a lognormal distribution. We do not observe a convergence of the higher statistics towards universal statistics, but do observe a strong temporal intermittency.

1. Introduction

We consider a straight line vortex imbedded in a three-dimensional periodic domain. We perturb the vortex and follow its evolution by a vortex method, in the hope that the calculation will shed light on aspects of the dynamics of vorticity which are significant for the understanding of turbulence.

The equations of motion are Euler's equations. The reasons for assuming that the viscosity is absent are spelled out in [3, 8]: It is reasonable and consistent with both numerical experience and available theory to assume that in a periodic domain the solution of the Navier-Stokes equations converge to the solution of the Euler equations strongly enough for the properties of the energy-containing and inertial ranges to be analyzable in the inviscid case. Such an assumption is implicitly made in Kolmogorov's theory of the inertial range.

The calculations can of course be pursued only for a short time, until the complexity of the flow outstrips the available computer memory and time. However, significant information can be gleaned in this short time. Long time calculations require a rescaling or a renormalization group procedure [8, 22, 29, 30].

---

* This work was supported in part by the Director, Office of Energy Research, Office of Basic Energy Sciences, Engineering, Mathematical and Geosciences Division of the U.S. Department of Energy, under contract W-7404-ENG-48, and in part by the Office of Naval Research, under contract N00014-76-C-0316

Reprinted with permission from *Communications in Mathematical Physics*, Vol. 83, 1982.

We obtain numerical results consistent with the conclusion reached in [8, 27] that the L_1 and L_2 norms of the vorticity become infinite in a finite time. We confirm the conclusion, reached in [8] by a rescaling procedure, that the L_2 support of the vorticity shrinks to an object of Hausdorff dimension ~ 2.5, as predicted by Mandelbrot [23, 24]. Graphical representations of the solution display the complexity of the flow and provide an intuitive explanation for the occurrence of strange sets.

The vortex motion presents an interesting mixture of coherence and disorder. Universal statistics are not achieved in our calculation, and support is found for the coherent structure model of the inertial range [3, 6]. The distribution of vorticity is seen to be approximately lognormal, in a sense different from both the Saffman model [28] and Kolmogorov's assumption [17]. A striking temporal intermittency is observed, somewhat similar to a phenomenon observed by Siggia [29].

In summary, the vorticity stretches wildly but the constraint of energy conservation prevents it from spreading evenly and forces it into tight knots. In general, the amount of disorder in a turbulent flow is presumably a function of the available energy.

2. The Equations of Motion and their Approximate Solution

The general framework for our calculation is similar to the one in [8]. The Euler equations for incompressible inviscid flow can be written in the form

$$\partial_t \boldsymbol{\xi} + (\mathbf{u} \cdot \nabla)\boldsymbol{\xi} - (\boldsymbol{\xi} \cdot \nabla)\mathbf{u} = 0, \tag{1a}$$

$$\boldsymbol{\xi} = \operatorname{curl} \mathbf{u}, \qquad \operatorname{div} \mathbf{u} = 0, \tag{1b, c}$$

where $\mathbf{u}$ is the velocity, $\boldsymbol{\xi}$ is the vorticity, t is the time, and ∇ is the differentiation vector. These equations are to be solved in a box of side 1, with periodic boundary conditions.

Suppose the initial data can be approximated by M vortex tubes of small but finite cross-section. The circulation of the ith tube is Γ_i. Let $\mathbf{r}(t)$ be the radius vector of a point moving with the fluid. The velocity induced by the tubes at $\mathbf{r}(t)$, as determined by Euler's equation, can be approximated by the Biot-Savart law (see [1]):

$$\mathbf{u}(\mathbf{r}) = -\frac{1}{4\pi} \sum_{i=1}^{M} \Gamma_i \int_{\substack{i^{\text{th}} \\ \text{tube}}} \frac{\mathbf{a} \times d\mathbf{s}}{a^3}, \tag{2}$$

where $\mathbf{s} = \mathbf{s}(\mathbf{r}')$ is the unit tangent to the ith tube at $\mathbf{r}'$, ds is the element of arc length along that axis, $d\mathbf{s} = \mathbf{s}\, ds$, $\mathbf{a} = \mathbf{r} - \mathbf{r}'$, and $a = |\mathbf{a}|$ is the length of $\mathbf{a}$. If $\mathbf{r}$ lies on one of the tubes, formula (2) has to be modified and the finite cross-section taken into account, because a can vanish and the velocity induced by close-neighbor interaction on infinitely thin tubes is, in general, infinite (see, e.g., [1]). Thus, if $\mathbf{r}(t)$ is a point on the axis of one of the tubes, its velocity is given approximately by

$$\mathbf{u}(\mathbf{r}) = -\frac{1}{4\pi} \sum_{i=1}^{M} \Gamma_i \int_{\substack{i^{\text{th}} \\ \text{tube}}} \frac{\mathbf{a} \times d\mathbf{s}}{\psi(a)}, \tag{3}$$

where $\psi(a) = a^3$ when a is large, and $\psi(a)$ satisfies, among other constraints, the condition

$$\lim_{\delta \to 0} \int_{\mathbf{r} - s\delta}^{\mathbf{r} + s\delta} \frac{\mathbf{a} \times d\mathbf{s}}{\psi(a)} = 0.$$

The motion of $\mathbf{r}$ is then given by

$$\frac{d\mathbf{r}}{dt} = \mathbf{u}(\mathbf{r}). \tag{4}$$

Equation (3) is our point of departure for approximating Eqs. (1). We assume that the initial data are such that the initial vorticity can be approximated by N vortex segments, i.e., short, thin, circular cylinders whose axis is tangent at a point to the vorticity vector (Fig. 1). The coordinates of the center of the base of the ith segment

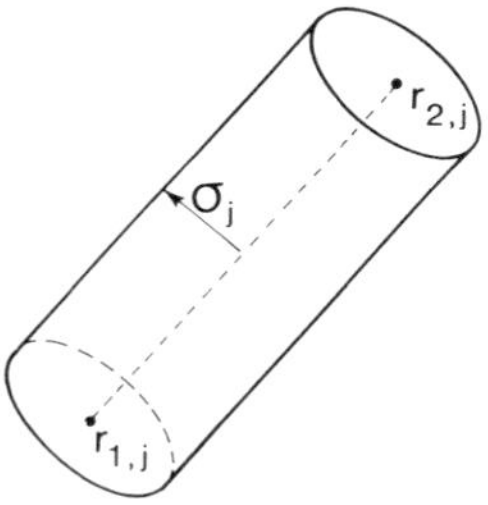

Fig. 1. A vortex segment

are $\mathbf{r}_i^{(1)} = (x_i^{(1)}, y_i^{(1)}, z_i^{(1)})$, and the coordinates of the center of the top are $\mathbf{r}_i^{(2)} = (x_i^{(2)}, y_i^{(2)}, z_i^{(2)})$. The ith segment has a "circulation" Γ_i,

$$\Gamma_i = \int_{\substack{\text{cross}\\ \text{section}}} \boldsymbol{\xi} \cdot \mathbf{d\Sigma}$$

and radius σ_i, $i = 1, \ldots, N$. Connected segments remain connected, $\mathbf{r}_i^{(2)} = \mathbf{r}_{i+1}^{(1)}$. (For another example of the use of such segments, see [7]; in the present inviscid calculation, the difference between an algorithm based on the use of segments and the filament algorithm of [8] is merely one of book keeping.) No segment is allowed to be longer than a predetermined small number h, $|\mathbf{r}_i^{(2)} - \mathbf{r}_i^{(1)}| \leqq h$ for all i.

The vectors $\mathbf{r}_i^{(1)}$, $\mathbf{r}_i^{(2)}$ move according to an approximation of Eq. (4):

$$\mathbf{r}_i^{(1)n} \equiv \mathbf{r}_i^{(1)}(nk), \quad \mathbf{r}_i^{(2)n} \equiv \mathbf{r}_i^{(2)}(nk), \quad n \text{ integer}, \; k = \text{time step};$$

$$\begin{aligned} \mathbf{r}_i^{(1)n+1} &= \mathbf{r}_i^{(1)n} + k\mathbf{u}_i^{(1)}, \\ \mathbf{u}_i^{(1)} &= -\frac{1}{4\pi} \sum_{j=1}^{N} \Gamma_j \frac{\mathbf{a} \times \Delta\mathbf{r}_j}{\phi(a)}, \\ \Delta\mathbf{r}_j &= \mathbf{r}_j^{(2)} - \mathbf{r}_j^{(1)}, \\ \mathbf{a} &= \tfrac{1}{2}(\mathbf{r}_j^{(2)} + \mathbf{r}_j^{(1)}) - \mathbf{r}_i^{(1)}, \\ a &= |\mathbf{a}|; \end{aligned} \tag{5}$$

with similar expressions for $\mathbf{r}_i^{(2)n+1}$. We choose the following form of ϕ [which corresponds to ψ in (3)]:

$$\frac{1}{\phi(a)} = \begin{cases} (a_{\min}^2 a)^{-1} & \text{if} \quad a \leqq a_{\min}, \\ a^{-3} & \text{if} \quad a_{\min} \leqq a \leqq a_{\max}, \\ 0 & \text{if} \quad a > a_{\max}, \end{cases}$$

$a_{\min}$, $a_{\max}$ are parameters to be chosen. The assumption $1/\phi = 0$ if $a > a_{\max}$ is convenient, and is reasonable since $1/\phi$ is small when a is large. If the vortex tubes are closed, there is no need to compute both $r_i^{(1)n+1}$ and $r_i^{(2)n+1}$. The time step k is chosen by requiring that

$$\max |\mathbf{r}_i^{(1)n+1} - \mathbf{r}_i^{(1)n}| = k \max |\mathbf{u}_i^{(1)n}| \leqq K, \qquad K = \text{small constant}. \tag{6}$$

There are thus four parameters: $a_{\min}$, $a_{\max}$, h and K to be chosen; this will be done in the next section. Note that in a periodic box each vortex element interacts not only with each other segment but also with an infinite set of images of that other segment; however, if $a_{\max} < \frac{1}{2}$ only one of these interactions is non-zero.

Note also that the function ϕ is independent of the curvature of the vortex tubes. Numerical methods with such geometry-independent cut-off functions ϕ have been tested, e.g., in [7, 8], and have been shown to converge in [2]. On the other hand, it is known that for a single vortex line the leading term in an asymptotic expansion of its induced velocity field is curvature-dependent (see, e.g., [1, 14, 19]). The paradox is resolved by the fact that a single physical vortex may have to be approximated by a cloud of vortex segments, whose collective motion resolves all effects, including local curvature effects. A similar situation holds for vortex motion in the plane, where clouds of non-deformable numerical vortex elements approximate well the motion of deformable physical vortices [13]. Furthermore, I have run three-dimensional calculations in which every vortex was moved only by its local curvature-dependent self-induction, following Hama [14]. The resulting motion turned out to be slow and the vortex stretching insignificant. This suggests that even though the self-induction curvature-dependent term may be large for a single vortex with an arbitrarily chosen geometry, in its natural motion a vortex rearranges itself so that the self-induction effect is lessened.

We shall assume that the vortex segments retain a cylindrical cross-section throughout their evolution, in the expectation that an arbitrary vorticity configuration can be approximated by cylindrical segments. The theory in [2] and the analogy with the two-dimensional situation lend support to this expectation. Furthermore, we assume that the distribution of vorticity remains uniform within each segment. This assumption is not essential, but does simplify the bookkeeping.

As the flow evolves, the vortex segments stretch. If a segment becomes longer than h, it is broken up into two segments, each with half the original length. The new end-points are determined by linear interpolation. The cumulative stretching of the line is tracked as follows: Each segment $\overline{\mathbf{r}_i^{(1)}\mathbf{r}_i^{(2)}}$ is assigned a tag q_i. When the ith segment is broken up into two halves, each one of the new segments is assigned a tag equal to $2q_i$. Initially, $q_i = l_{0i}/V_i$, where V_i is the volume of the ith segment and l_{0i} is its initial length. The number of segments, N, increases with time. In our calculations, we assumed that initially $V_i = V_0$ for all i; we also assumed that the segments had equal initial lengths $l_0 \leqq h$.

Let l_i be the length of the ith segment, $l_i = |\mathbf{r}_i^{(2)} - \mathbf{r}_i^{(1)}|$, $|l_i| \leqq h$. The tag assigned to that segment had been doubled each time the length of a segment had been halved. The cross-section of the segment is thus $l_0/l_i q_i$, and its volume is l_0/q_i. Thus, the total volume occupied by the vortex segments is

$$V = l_0 \sum_{i=1}^{N} \frac{1}{q_i};$$

it is easy to see that V is constant in time with our computing scheme, as Eq. (1c) requires.

Assume that initially each vortex segment contains vorticity $\boldsymbol{\xi}$, parallel to its axis, with $|\boldsymbol{\xi}| = \xi_0$ equal for all segments. The L_1 norm of the vorticity, $\|\boldsymbol{\xi}\|_1 = \int |\boldsymbol{\xi}|\, dV$, can be evaluated as follows: Since the cross-section of the ith segment is $l_0/l_i q_i$, the vorticity in the ith segment is proportional to $l_i q_i$, and, therefore, up to an immaterial constant,

$$\|\boldsymbol{\xi}\|_1 = \sum_{i=1}^{N} l_i. \tag{7}$$

By a similar argument, the L_2 norm of $\boldsymbol{\xi}$ equals

$$\|\boldsymbol{\xi}\|_2^2 \equiv \int |\boldsymbol{\xi}|^2 dV = \sum_{i=1}^{N} q_i l_i^2, \tag{8}$$

where, again, an immaterial constant has been omitted. For simplicity, we shall henceforth omit factors such as l_0 when they play no role.

All the calculations below were made with the following initial data: A vertical cylindrical vortex is deformed in such a way that its axis consists of four straight lines through the points $(\frac{1}{2}, \frac{1}{2}, 0)$, $(\frac{1}{2}, \frac{1}{2}, \frac{2}{5})$, $(\frac{1}{2}, \frac{1}{2} + 0.1, \frac{3}{5})$, $(\frac{1}{2}, \frac{1}{2}, \frac{4}{5})$, $(\frac{1}{2}, \frac{1}{2}, 1)$. This vortex is then divided into segments of length smaller than h, and is assigned some cross-section S_0. $\Gamma_j = 1$ for all j.

Note that in [7] we set $\sigma_i = a_{\min}$ for all i. This identification was natural, but in no way logically required, and will not be used here.

3. Accuracy and Numerical Parameters

In the present section we show how the numerical parameters needed in our calculation are picked, and demonstrate that under suitable conditions the results obtained are independent of these parameters.

Consider first the dependence of the computed solution on the parameter K which determines the time step k. The behavior of the solution as a function of time is hard to use in assessing accuracy, since the velocity $\mathbf{u}$ increases very fast and a substantial part of the total time elapsed in 40 time steps is in fact spent in the first time step. Since the initial data are fixed, the length of the first step is proportional to K, and solutions with different K's appear as translates of each other in time. In

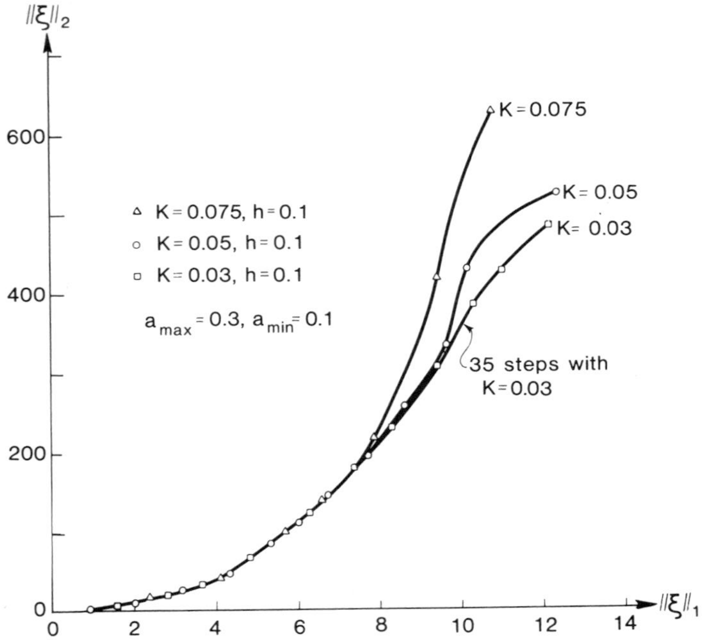

Fig. 2. (L_1, L_2) portrait of the calculations

order to remove this effect, we consider the (L_1, L_2) portrait of the flow, i.e., the curve traced by the flow in a plane where the coordinates are $\|\xi\|_1$ and $\|\xi\|_2$. From formulas (7), (8) it is seen that $\|\xi\|_2$ depends more on the total stretching than does $\|\xi\|_1$.

In Fig. 2 we display the (L_1, L_2) portraits of calculations performed with $a_{\min}=0.1$, $a_{\max}=0.3$, $h=0.1$, and $K=0.075$, 0.05, 0.03. One can see that as $K\to 0$ the curves converge to a limiting curve. A calculation with $K=0.02$ is indistinguishable in this representation from the calculation with $K=0.03$, for the times under consideration. When $a_{\min}=0.1$, $a_{\max}=0.3$, $h=0.1$, we choose $K=0.03$.

The runs made with $h=0.075$ and $h=0.05$ ($a_{\min}=0.1$, $a_{\max}=0.3$, $K=0.03$) are indistinguishable in the (L_1, L_2) plane from the one made with $h=0.1$. The runs made with $h=0.1$, $K=0.03$, $a_{\min}=0.1$, and $a_{\max}=0.2$, 0.3, 0.4, 0.5 are also indistinguishable.

However, if $a_{\min}$ is reduced, K and h have to be reduced also. When they are reduced substantially the calculation returns to the portrait it had with $a_{\min}=0.1$. One can see that all that happens is that the time and space scales are reduced and the calculation merely rescaled. Thus, we shall pick in all the runs below $a_{\min}=0.1$, $a_{\max}=0.3$, $h=0.1$, $K=0.03$.

We also made some runs with $a_{\min}$ variable, and dependent on the local cross-section of the segments. With appropriate values of K, $a_{\max}$ and h the calculation gives the same results as the one with constant $a_{\min}$.

The lack of sensitivity of the results to parameters such as $a_{\min}$, $a_{\max}$ underscores, once again, the fact that some of the properties of turbulence are reasonably independent of the exact form of the equations of motion, a phenomenon already familiar from the theory of critical phenomena (see, e.g., [30]).

Another check on the accuracy of the calculation is the verification of energy conservation. The energy in the periodic box is

$$E = \tfrac{1}{2} \int |\mathbf{u}|^2 \, dV. \quad \mathbf{u} = \text{velocity vector}.$$

E should be approximately constant in time. $\mathbf{u}$ can be computed at an arbitrary point by an obvious extension of formulas (5). The flow is very complex (see below) and it is hopeless to try to evaluate E by classical quadrature. The best we can do is distribute some points evenly in the box, evaluate $|\mathbf{u}|^2$ at these points, and average; in Table 1 we display the results of two such calculations, with $K = 0.03$, $a_{\min} = 0.1$, $a_{\max} = 0.3$, $h = 0.1$, and with 5^3 and 9^3 sample points respectively. Energy does appear to be as constant as the method by which it is evaluated would allow. In Table 2 we display the results of a run made with $K = 0.1$ (a value which we already know is too large). Energy is not conserved.

Each one of the major calculations below has been checked to see how sensitive it was to a variation in the numerical parameters.

Table 1. Energy conservation, $K = 0.03$

Step	5^3 sample points	9^3 sample points
1	0.17	0.19
5	0.16	0.18
10	0.15	0.17
15	0.16	0.16
20	0.18	0.15
25	0.18	0.17
30	0.20	0.18

Table 2. Energy (non)conservation, $K = 0.1$. 9^3 sample points

Step	Energy
1	0.14
5	0.15
10	0.21
12	0.77
14	0.32
16	0.45

4. Main Features of the Flow

As soon as the flow begins, the vorticity begins to stretch, and the stretching is extraordinarily rapid. In Fig. 3 we plot the evolution of $\|\xi\|_1$ as a function of time. The graph is consistent with the conclusion in [8] that $\|\xi\|_1$ (and *a fortiori* $\|\xi\|_2$)

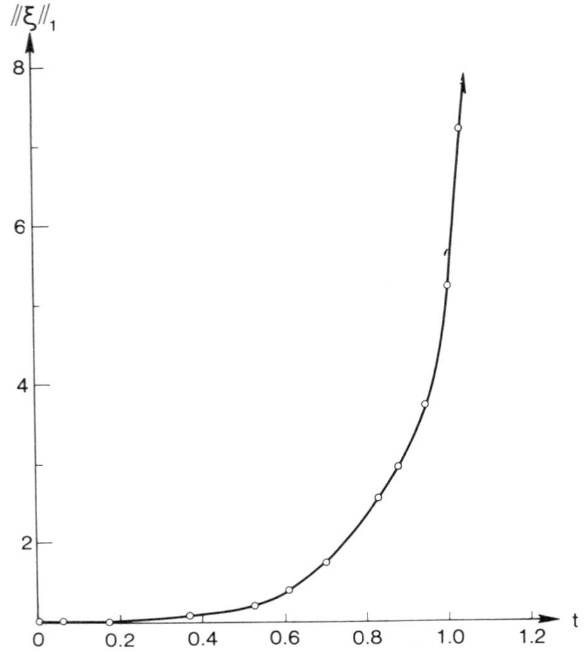

Fig. 3. $\|\xi\|_1$ as a function of time

become infinite in a finite time. Of course, as $\|\xi\|_1$ increases, the number of segments needed in the calculation increases, and the calculation cannot be continued forever.

Attach to each segment a stretching number $s_i = q_i l_i$, where q_i is the tag attached to the vortex segment and l_i is its length, $l_i = |\mathbf{r}_i^{(2)} - \mathbf{r}_i^{(1)}| \leqq h$. The length scale associated with each vortex is the square root of its cross-section; that scale is proportional to $1/\sqrt{s_i}$. By step 40, some segments have been stretched by a factor of ~ 400, while others have been stretched only by a factor of ~ 2. The ratio of the largest to the smallest scale at the end of step 40 is thus about 20 to 1. This is not a large enough spread to allow a determination of the inertial range exponent. We tried to calculate the average value of $(\mathbf{u}(\mathbf{x}+\mathbf{r}) - \mathbf{u}(\mathbf{x}))^2$, where $\mathbf{u}$ is the velocity and $r = |\mathbf{r}|$ is comparable with the scales present in the calculation, and then approximate this function by r^γ. The values of γ obtained in this way were not independent of the range of scales chosen, and ranged between 0.9 and 1.4.

As the flow evolves, $u_{\max} = \max_i |\mathbf{u}_i^{(1)}|$ increases rapidly even though the energy remains constant, and k, the time step, decreases from 0.37 at step 1 to 0.013 at step 40. An interesting interpretation of that fact is presented below.

In Fig. 4 we present the general configuration of the vortex segments after 10 steps ($t = 0.65$), 20 steps ($t = 0.88$), 30 steps ($t = 1.04$) and 40 steps ($t = 1.21$). The pattern of increasing complexity is obvious. However, the vortex does not forget the fact that its initial configuration was vertical (for a quantitative discussion, see below). Some entrainment of irrotational fluid does take place, and can be measured as follows: Define the horizontal center of gravity of the vortex,

$$\bar{x} = \frac{1}{N} \sum_{i=1}^{N} x_i^{(1)}, \qquad \bar{y} = \frac{1}{N} \sum_{i=1}^{N} y_i^{(1)}.$$

The entrainment radius R is

$$R = \frac{1}{N} \sum_{i=1}^{N} \sqrt{(x_i^{(1)} - \bar{x})^2 + (y_i^{(1)} - \bar{y})^2}. \tag{9}$$

R increases from $R = 0.025$ at $t = 0.37$ to $R = 0.137$ at $t = 1.17$.

The most remarkable feature of Fig. 4 is the fact that as the vortices stretch they organize themselves into coherent narrow sheaves. In Fig. 5 we show details of the structure at step 28, i.e., relatively early. At later times, the packing of the segments is so tight that it is difficult to discern in them a consistent pattern. (Indeed, one would not expect to have an intuitive grasp of an object of Hausdorff dimension ~ 2.5.) Some of the "legs" in Fig. 4d contain 20 to 30 separate segments.

The explanation for this phenomenon is as follows: As the vortices stretch, their cross-section decreases and the energy associated with them would increase unless they arranged themselves in such a way that their velocity fields cancelled. The folding achieves such cancellation; it will be reinterpreted in the next section in terms of Hausdorff dimension.

The large degree of coherence in the physical vortex cores, and the fact that stretching and folding rearrange the vorticity in thin, well-defined vortex structures, provide some support to the conjecture in [3, 6] that the inertial range spectrum is related to the spectral tail of the vorticity distribution in discrete vortex structures.

5. Hausdorff Dimension of the Support of Vorticity

In [8] we presented a calculation which verified Mandelbrot's conjecture that the L_2 support of the vorticity shrinks into a set of Hausdorff dimension ~ 2.5. The calculation in [8] was based on a rescaling argument whose validity is not rigorously established (for an example where rescaling has been applied and can be proved to be valid, see [9]). In the present section we obtain the same result by a different method.

We first define Hausdorff dimension. Consider a compact set C; cover it by a finite collection of balls of radii ϱ_i, $\varrho_i \leqq \varrho$. Form the sum

$$S(D) = \sum_i \varrho_i^D, \qquad D = \text{positive number}.$$

Consider the quantity

$$h(D) = \lim_{\varrho \to 0} \liminf S(D).$$

$h(D)$ is the Hausdorff measure of C in dimension D. $h(D)$ is zero for D large, usually infinite for D small; the number

$$D^* = \begin{cases} \text{greatest lower bound of } D \text{ for which } h(D) \text{ is zero,} \\ \text{smallest upper bound of } D \text{ for which } h(D) \text{ is infinite,} \end{cases}$$

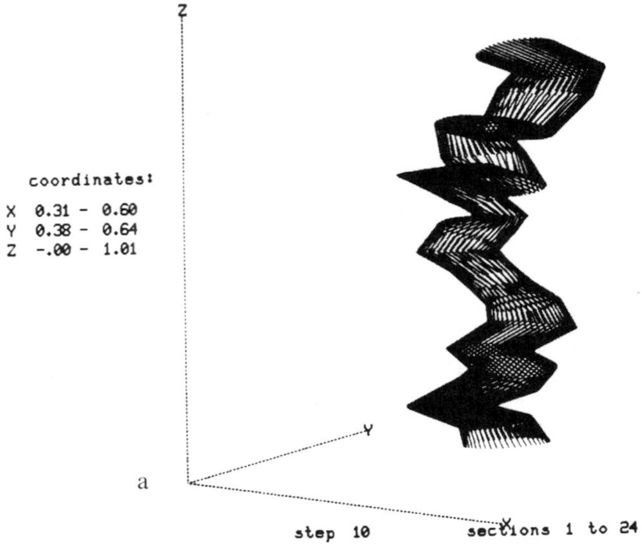

a

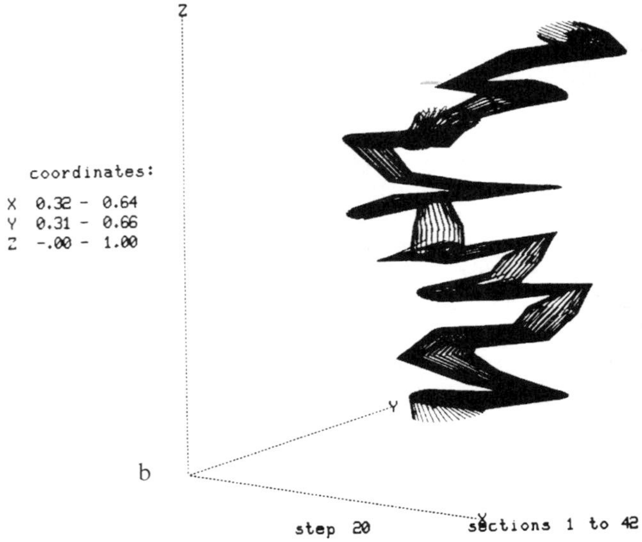

b

Fig. 4a–d. The evolution of the vortex: **a** step 10, time = 0.65 **b** step 20, time = 0.88

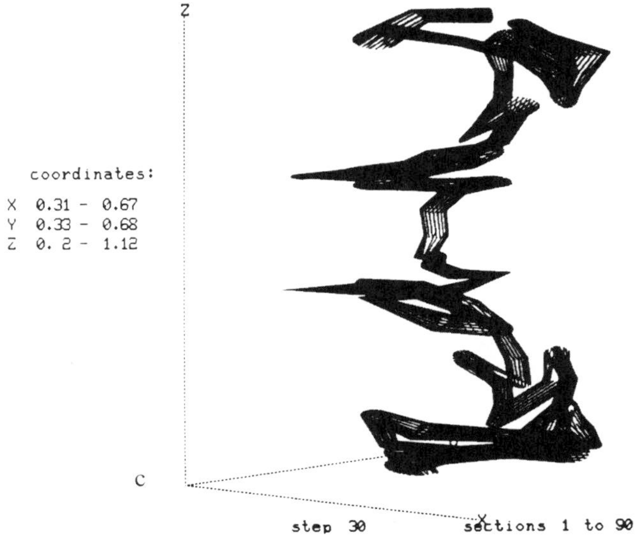

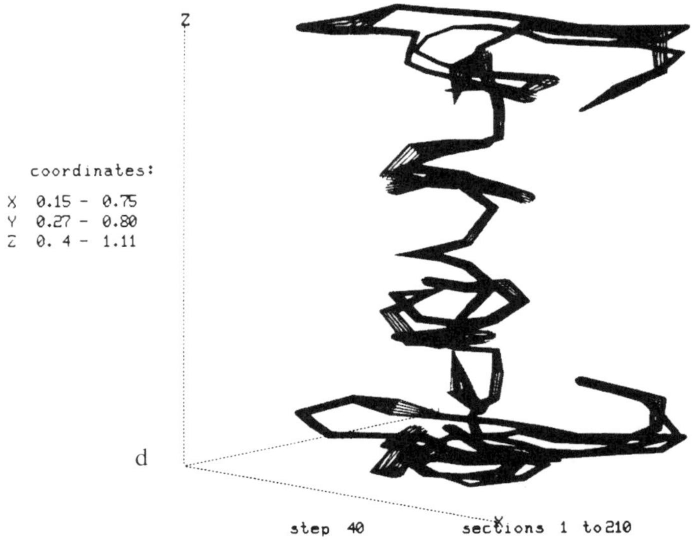

c step 30, time = 1.04 **d** step 40, time = 1.21

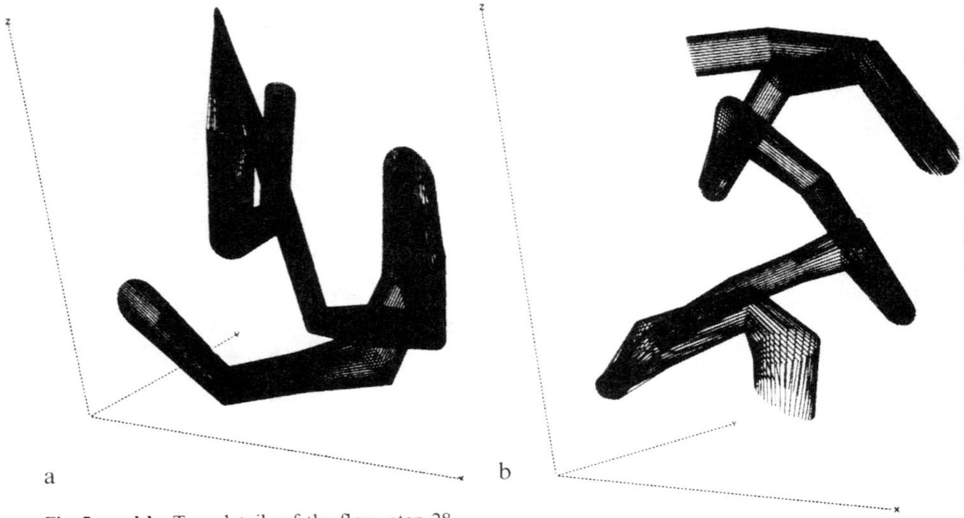

Fig. 5a and b. Two details of the flow, step 28

is the Hausdorff dimension of C. For a cube $D^*=3$, for a square $D^*=2$, for a line segment $D^*=1$, for the tertiary Cantor set $D^*=\log 2/\log 3$ (see [16]). The balls used in the definition of D^* can be replaced by any family of non-degenerate self-similar objects [12].

The conjecture verified in [8] is the following: Consider $\|\xi\|_2^2=\int|\boldsymbol{\xi}|^2 dV$. Consider any $\varepsilon>0$. There is a time T such that for $t>T$ $\boldsymbol{\xi}=\boldsymbol{\xi}_1+\boldsymbol{\xi}_2$, the supports of $\boldsymbol{\xi}_1$ and $\boldsymbol{\xi}_2$ are disjoint, the support of $\boldsymbol{\xi}_1$ has dimension $D^*\cong 2.5$, and $\|\boldsymbol{\xi}_1\|_2^2=(1-\varepsilon)\|\boldsymbol{\xi}\|_2^2$; i.e., almost all the vorticity has support of dimension D^*.

We verify this conjecture here by a different method. Consider the stretching numbers $s_i=l_i q_i$. We pick an ε, and determine $s^*=s^*(\varepsilon)$ such that

$$\sum_{\substack{\text{segments}\\ \text{such that}\\ s_i>s^*}} q_i l_i^2=(1-\varepsilon)\sum_{\substack{\text{all}\\ \text{segments}}} q_i l_i^2,$$

[see formula (8)]. We cover the segments for which $s_i>s$ by cylinders with circular bases whose heights are equal to the radius ϱ_i of their bases. A segment of length l_i with tag q_i has cross-section $1/l_i q_i$ and can be covered by cylinders with $\varrho_i=1/\sqrt{l_i q_i}$. The sum $S(D)$ which corresponds to this cover is

$$S(D)=\sum_{s_i>s^*} l_i\sqrt{l_i q_i}\left(\frac{1}{l_i q_i}\right)^{D/2},\qquad s^*=s^*(\varepsilon). \tag{10}$$

$S(D)$ approximates the lim inf of sums corresponding to covers with such cylinders. Indeed, consider one segment. A cover with smaller cylinders will only increase $S(D)$ (since when we double the number of cylinders we decrease the factor ϱ^D by less than two); on the other hand, if we cover the segment by larger cylinders of radii, say, α, there will be $\sim l_i/\alpha$ such cylinders and the corresponding contribution

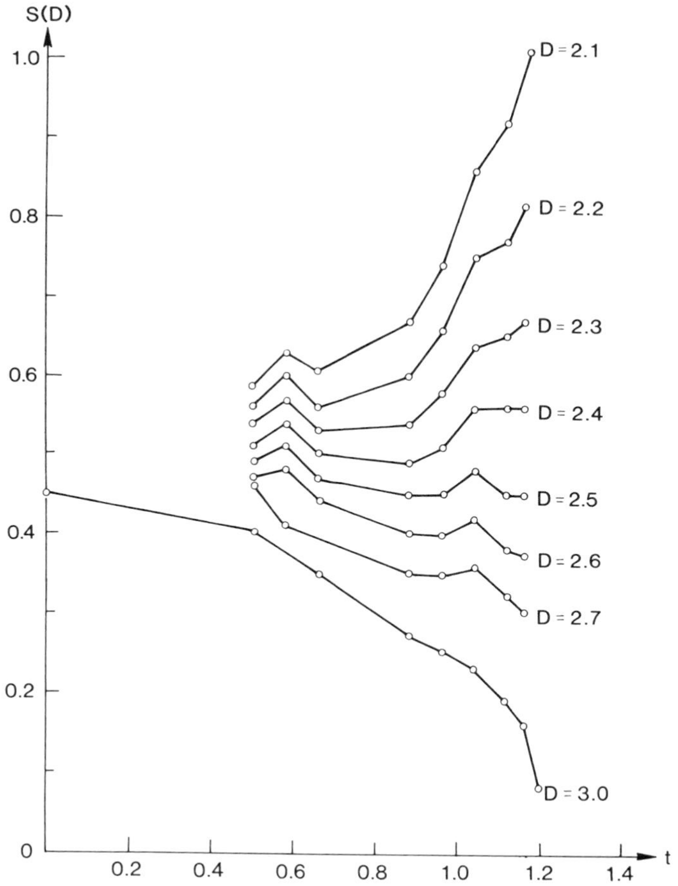

Fig. 6. Hausdorff sums as functions of time

to $S(D)$ will be $\sim (l_i/\alpha)\alpha^D = l_i \alpha^{D-1}$. If $D > 1$ (which will turn out to be the case in all the calculations below) the contribution to $S(D)$ is smallest when α is as small as possible. We follow the sums $S(D)$ in time, relying on the fact that the segments are stretching, and thus that the $1/\sqrt{l_i q_i}$ are decreasing, to produce the limit of vanishing linear dimension.

In Fig. 6 we plot the evolution of $S(D)$ for several values of D, with $\varepsilon = 0.1$. For small t, $S(D)$ is not always defined because there are not enough segments for s^* to be defined. Note that at $t = 1.21$, $S(3) = 0.08$; initially $S(3) = 0.45$; i.e., at $t = 1.21$, 90% of the squared vorticity is contributed by 16% of the volume originally occupied by vorticity. For $D < D^*$, $S(D)$ should be increasing; for $D > D^*$, $S(D)$ should be decreasing. An inspection of Fig. 6 leads to the estimate $D^* \sim 2.5$, in agreement with the conclusion of [8].

The notion of Hausdorff dimension provides an interpretation of the process of vortex folding described in the preceding section. The vorticity keeps on stretching and eventually, after a finite or infinite time has elapsed, some cross-section of a vortex line should have zero Lebesgue measure. However, that cross-section cannot shrink to a point because the energy associated with a point vortex is infinite. Sets of non-integer Hausdorff dimension appear in problems where, for example, one tries to characterize sets of zero Lebesgue measure which can carry a finite charge while giving rise to a bounded potential (see [12]). It seems likely that the constraint of finite energy requires that a cross-section of an infinitely stretched vortex have a sufficiently large Hausdorff dimension. The cross-section of a set of Hausdorff dimension $2+\alpha$ will in general have Hausdorff dimension $1+\alpha$ (see [26]). The process of vortex folding is the process by which such a cross-section is generated.

In [11], Frisch et al. presented an interesting derivation of the relationship between an energy cascade and Hausdorff dimension. They considered eddies of typical linear dimension L, containing energy proportional to βU^2, where β is the decreasing fraction of volume occupied by active eddies of linear dimension L. The characteristic time T of such eddies is $\sim L/U$, and the Kolmogorov assumption is that in a characteristic time T the eddies lose their energy to smaller eddies, the rate of energy transfer $\beta U^2/T = \beta U^3/L$ being constant. It is difficult to verify such assumptions on the computer, since quantities such as U, L and T are not sharply defined. However, consider the flow in the periodic cube as making up a single eddy. $\beta U^2 =$ constant, and if β is decreasing, U should increase, and it does. The quantity $\max_i |u_i|$ increases while $\int u^2\, dV$ remains constant because the activity is confined to an ever decreasing volume. Accuracy requires that the time step k be a small fraction of the characteristic time. The characteristic time. L/U, is decreasing, and thus k should decrease; we have seen that it does.

6. Lognormality of the Vorticity Distribution

As the vortex lines stretch, the range of values $s_i = q_i l_i$, $i = 1, \ldots, N$, assumed by the vorticity increases. It is of interest to determine the distribution of the s_i.

In [2], Saffman provided an argument to show that the distribution of values assumed by the vorticity is lognormal; he assumed that the local rate of stretching is proportional to the local vorticity multiplied by a random coefficient:

$$\frac{d\xi}{dt} = b(t)\,\xi. \tag{11}$$

By integration we find

$$\log(\xi(t)) - \log(\xi(0)) = \int_0^t b(t)\,dt,$$

and if the values of $b(t)$ for distinct values of t are reasonably independent, and if all the $\log \xi(0)$ are equal, it follows that the distribution of $\xi(t)$ is lognormal. Assumption (11) is reasonable for a short time, for indeed the more vorticity has been stretched in a neighborhood, the more vorticity is available to perform further stretching; the stretching also depends on the geometrical configuration of the

Table 3. Skewness and flatness of log s

Step	Skewness	Flatness
1	0.40	1.16
5	0.77	2.17
10	0.74	2.22
15	0.54	2.43
20	0.34	2.59
25	−0.60	3.35
30	−0.29	2.39
35	−0.19	2.25
40	−0.19	2.81

vortex which can quite reasonably be viewed as random. However, once the vorticity has been stretched a lot, vorticity contributions generated in one part of the flow interact with vorticity contributions generated in another part of the flow, so that (11) is no longer a convincing model.

A different lognormality was assumed to hold in Kolmogorov's theory [17]. He assumed that the distribution of dissipation in disjoint, fixed small volumes is lognormal. The two lognormality assumptions are different, and, as shown in [18] and [23], probably incompatible. We have been able to verify neither.

However, if we consider the distribution of the s_i's obtained by our algorithm, we see that it does at least approximately converge to a lognormal distribution. In Table 3 we display the skewness and flatness of the distribution of the computed $\log s_i$ as functions of time. These quantities are defined as follows: Let α be a random variable, and let an overbar denote an average at a fixed time. $V(\alpha)=\overline{(\alpha-\bar{\alpha})^2}$ is the variance of α, $Z(\alpha)=\overline{(\alpha-\bar{\alpha})^3}/V(\alpha)^{3/2}$ is the (normalized) skewness of α, and $K(\alpha)=\overline{(\alpha-\bar{\alpha})^4}/(V(\alpha))^2$ is the flatness of α. For a gaussian variable α, $Z(\alpha)=0$ and $K(\alpha)=3$. The values of $Z(\log s)$ and $K(\log s)$ have an error of approximately ± 0.3 (as can be seen by making several runs with different numerical parameters) and are compatible with the conclusion that $\log s$ has a normal distribution. In Fig. 7 we display the distribution of $(\log s-\overline{\log s})/V(\log s)$

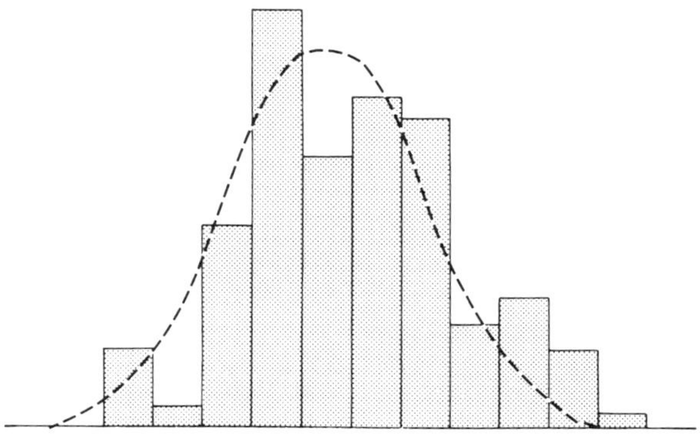

Fig. 7. Histogram of the vorticity distribution

as well as the gaussian distribution with the same mean and variance, after the 40th step. In view of the fairly small sample size, the assumption that $\log s$ is gaussian becomes quite tenable.

This conclusion agrees with Saffman's conjecture if the total amount of stretching is small, but differs from Saffman's conjecture when the stretching is substantial. Indeed, in our calculation each vortex is divided into shorter vortices when its length exceeds h, and each new piece contributes a value of $\log s$ when the statistics of $\log s$ are computed. On the other hand, in Saffman's model, each vortex contributes a single value however much it may have been stretched. Our distribution is therefore much less "intermittent" than Saffman's, and allows fewer extreme values of s. Our conclusion would agree with Kolmogorov's if the segments were equidistributed in space and if one could apply his conjecture to volumes which contain exactly one segment; the first condition is unlikely to hold (see Fig. 4) and the second condition is not compatible with the analysis in [17].

7. Temporal Intermittency and Higher Statistics

It has already been mentioned that during the short time interval for which our problem can be run, the flow does not forget its initial data; in particular, the velocity field remains on the whole the velocity field of a vertical vortex. Let $\mathbf{u} = (u, v, w)$ be the velocity vector. Quantities such as $\overline{u^2}$, $\overline{v^2}$, $\overline{w^2}$, where the overbar denotes spatial averages, can be computed by the sampling method described earlier for computing energy. At $t=0$ $\overline{w^2} \sim 0$; $\overline{w^2}$ increases slowly (to about 20% of $2E = \overline{u^2} + \overline{v^2} + \overline{w^2}$) and then starts to decrease again.

In Fig. 8 we display $\overline{u^2}$, $\overline{v^2}$, $\overline{w^2}$ averaged over the central subregion $C_R\colon \frac{1}{2} \pm R \leqq x \leqq \frac{1}{2} + R$, $\frac{1}{2} \pm R \leqq y \leqq \frac{1}{2} + R$, $0 \leqq z \leqq 1$, where R is the entrainment radius defined in Eq. (9). We picked that region because the fluid can be viewed as more fully turbulent there and one could have expected a closer approximation to energy equipartition between u, v, w in that region than in the cube as a whole. However, just the opposite is the case. $\overline{u^2}$ and $\overline{v^2}$ oscillate, as one would expect from the fact that the vortex as a whole precesses as a consequence of its initial perturbation; $\overline{w^2}$ increases sharply but then decreases sharply, leaving an almost two-dimensional flow. The graph of $\overline{w^2}$ contains a sharp blip.

It was interesting to see if a similar blip could be seen in any other statistical description of the flow. None could be seen in quantities such as the flatness or the skewness of the distribution of u or v (which remain roughly constant and merely reflect the fact that we have approximately a velocity field associated with a vertical vortex). We therefore tried to compute the skewness and flatness of velocity derivatives such as u_x.

u_x can be computed formally from (5) by differentiation:

$$u_x(\mathbf{r}) = C \sum_{j=1}^{N} (\mathbf{a} \times \Delta\mathbf{r}_j)_1 / \tilde{\phi},$$

where

$$\mathbf{a} = \tfrac{1}{2}(\mathbf{r}_j^{(2)} + \mathbf{r}_j^{(1)}) - \mathbf{r},$$

$$a = |\mathbf{a}|,$$

$$\Delta\mathbf{r}_j = \mathbf{r}_j^{(2)} - \mathbf{r}_j^{(1)}.$$

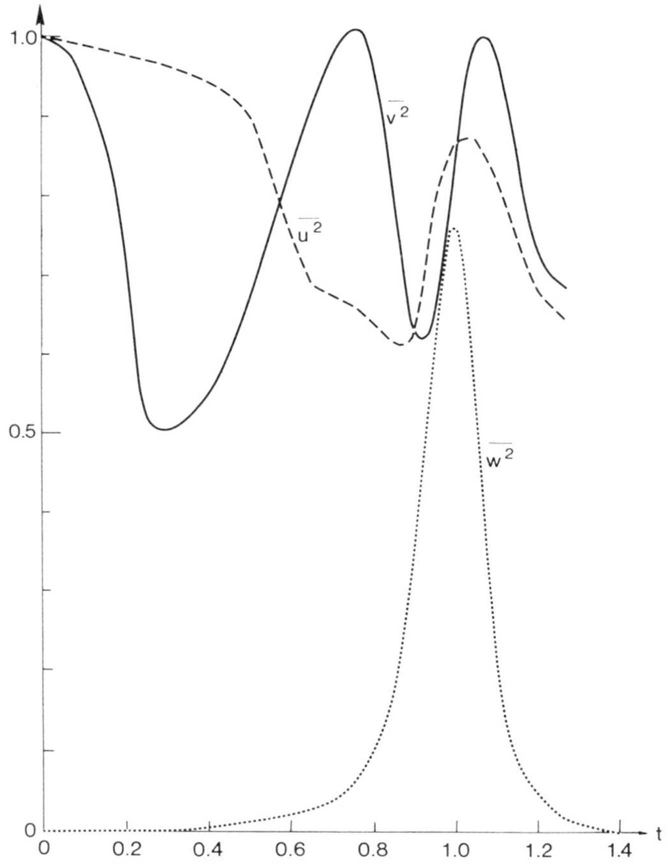

Fig. 8. $\overline{u^2}$, $\overline{v^2}$, $\overline{w^2}$ as function of time

$(\mathbf{a} \times \mathbf{r}_j)_1$ denotes the x component of $\mathbf{a} \times \Delta\mathbf{r}_j$, and

$$1/\tilde{\phi} = 1/\tilde{\phi}(a) = \begin{cases} a^{-2} a_{\min}^{-2} & \text{if} \quad a \leqq a_{\min} \\ a^{-4} & \text{if} \quad a_{\min} \leqq a \leqq a_{\max} \\ 0 & \text{if} \quad a > a_{\max} . \end{cases}$$

The constant C incorporates the factors $-1/4\pi$ and Γ_j of (5) as well as numerical coefficients which arise in the differentiation. It is not at all obvious that this differentiation leads to an approximation of u_x, and indeed the moments of u_x depend on $a_{\min}$, $a_{\max}$ as well as on the size of the region in which they are evaluated; only their qualitative behavior is of possible significance. The flatness of u_x exceeds the flatness of u for all choices of parameters.

It is not at all obvious either that the moments of u_x remain bounded in time (and indeed, we claimed earlier that $\|\boldsymbol{\xi}\|_2$ did not remain bounded). Also, the theorem on equality of Hausdorff dimension and capacitory dimension [12]

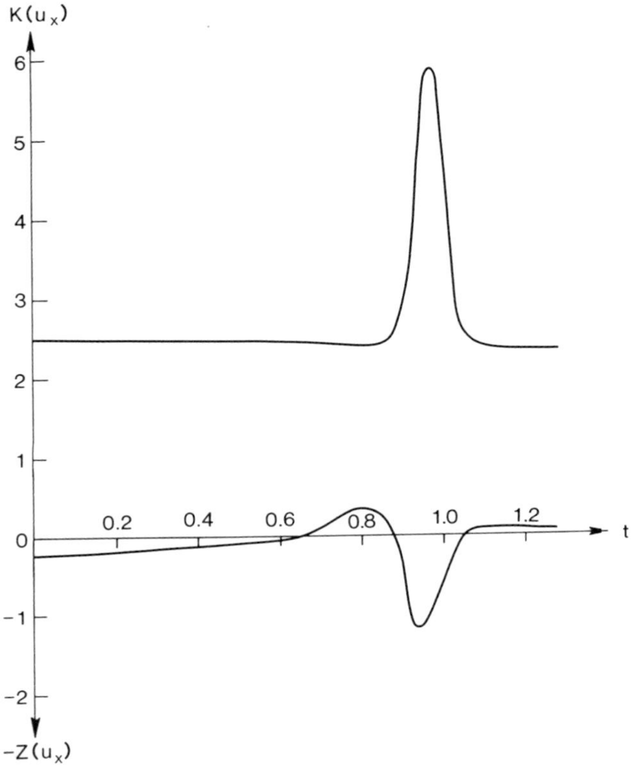

Fig. 9. Qualitative behavior of the skewness and flatmess of u_x

suggests that moments of sufficiently high derivatives of u do not exist, a conclusion reached by other means in [25]; this possibility casts a further doubt on the validity of the calculation of the moments of u_x.

Be that as it may, we have plotted in Fig. 9 the behavior of the skewness $Z(u_x)$ and flatness $K(u_x)$ of u_x averaged in space over the central region C_R. The temporal blip observed in the evolution of $\overline{w^2}$ has its clear equivalent here. Changes in the region over which the average is taken and in $a_{\min}$, $a_{\max}$ change the numerical values of S and K, but do not change the shape of the curve. The sudden increase in activity associated with the blip resembles the temporal intermittency observed by Siggia [29]. We have no good explanation for this phenomenon. It may be due to a vortex breakdown, such as the one observed numerically in [10], which could be responsible for the horizontal loops in Fig. 4d.

Note that nowhere in Figs. 8 and 9 do we observe a convergence towards statistics independent of the initial data. This may be due to the fact that the integration time is too short, but it could also be due to the non-existence of "universal" statistics. The experimental data (see, e.g., [31]) do not rule out the latter hypothesis. Arguments for and against the existence of "universal" statistics can be found e.g. in [25] and [29]. It seems likely that "universal" values of skewness and flatness, if they exist, depend on temporal as well as spatial averaging. The calculation of β in [8] lends support to this conjecture.

8. Conclusion

We have provided quantitative and qualitative information about the evolution of a three-dimensional vortex, which has a substantial bearing on the assessment of various theoretical models of turbulent flow. Most importantly, we have demonstrated the eminent suitability of vortex methods for the analysis of turbulence. Long thin objects which arise in fluid turbulence are easier to represent as long thin objects than in any other way.

The calculations above were performed on a VAX computer at the Lawrence Berkeley Laboratory. Listings of the programs used are available from the author.

References

1. Batchelor, G.K.: An introduction to fluid mechanics. Cambridge: Cambridge University Press 1967
2. Beale, T., Majda, A.: Vortex methods: convergence in three dimensions. Math. Comp (to appear)
3. Chorin, A.J.: In Proc. 2d Int. Conf. Num. Meth. Fluid Mech., pp. 285–289 Berlin, Heidelberg, New York: Springer 1970
4. Chorin, A.J.: J. Fluid Mech. **57,** 785–796 (1973)
5. Chorin, A.J.: J. Fluid Mech. **63,** 21–32 (1974)
6. Chorin, A.J.: Lectures on turbulence theory. Boston: Publish/Perish 1975
7. Chorin, A.J.: SIAM J. Sc. Stat. Comp. **1,** 1–24 (1980)
8. Chorin, A.J.: Comm. Pure. Appl. Math. **34,** 853–866 (1981)
9. Chorin, A.J.: Numerical estimates of Hausdorff dimension. J. Comp. Phys. (to appear)
10. del Prete, V.: Numerical simulation of vortex breakdown. LBL Math. & Computing Report, Berkeley (1978)
11. Frisch, U., Salem, P.L., Nelkin, M.: J. Fluid Mech. **87,** 719–736 (1978)
12. Frostman, O.: Potentiel d'equilibre et capacité des ensembles. Thesis Lund (1935)
13. Hald, O.: SIAM J. Num. Anal. **10,** 726–755 (1979)
14. Hama, F.: Phys. Fluids **6,** 526–528 (1963)
15. Hausdorff, F.: Dimension und äußeres Maß. Math. Ann. **79,** 1–21 (1919)
16. Kahane, J.P., Salem, R.: Ensembles parfaits et séries trigonométriques. Paris: Hermann 1963
17. Kolmogorov, A.N.: J. Fluid Mech. **13,** 82–86 (1962)
18. Kraichnan, R.H.: J. Fluid Mech. **62,** 305–330 (1974)
19. Lamb, H.: Hydrodynamics. London: Dover 1960
20. Leonard, A.: Proc. 4th Int. Conf. Num. Meth. Fluid Mech., pp. 245–250. Berlin, Heidelberg, New York: Springer 1975
21. Leonard, A.: J. Comput. Phys. **37,** 289–335 (1980)
22. Ma, S.K.: Modern theory of critical phenomena. Reading: Benjamin, 1977
23. Mandelbrot, B.: In: Statistical models and turbulence, Rosenblatt, M., Van Atta, C. (eds.) pp. 333–358. Berlin, Heidelberg, New York: Springer 1972
24. Mandelbrot, B.: J. Fluid Mech. **62,** 331–358 (1974)
25. Mandelbrot, B.: In: Turbulence and NS equation, Temam, R. (ed.) pp. 121–145. Berlin, Heidelberg, New York: Springer 1976
26. Mattila, P.: Ann. Acad. Sci. Finn. **A,** 1 (1975)
27. Morf, R., Orszag, S., Frisch, U.: Phys. Rev. Lett. **44,** 572–575 (1980)
28. Saffman, P.: Phys. Fluids **13,** 2193–2194 (1970)
29. Siggia, E.: J. Fluid Mech. **107,** 375–406 (1981)
30. Salem, P.L., Fournier, J.D., Pouquet, A.: In: Dynamical critical phenomena and related topics, Enz, C. (ed.). Berlin, Heidelberg, New York: Springer 1979
31. Van Atta, C., Antonia, R.: Phys. Fluids **23,** 252–268 (1980)

Communicated by J. Glimm

Received September 25, 1981